Advances in Integrated Pest and Disease Management in Horticultural Crops

Advances in Integrated Pest and Disease Management in Horticultural Crops

Vol. 4

Dr. P. Singh

AGROTECH PRESS

JAIPUR ▪ NEW DELHI

AGROTECH PRESS

Regd. Off. "Bony Residency", Gate No. 2
Opp. Tilak Public School, Vishweswariya Nagar,
Gopalpura Road, Jaipur - 302018, India
Ph: 0141 - 2761280, Telefax: 2761381
E-mail : agrotechpress@gmail.com

Delhi Office: 102, 1st Floor, Satyam House,
4327/3, Ansari Road, Darya Ganj,
New Delhi - 110002, India
Ph: 011-45652440
visit us: www.agrotechpress.com

Distributed By
KEYA INTERNATIONAL PRESS LLP
E-mail: keyainternationalpressllp@gmail.com

Published in India, 2024

ISBN : 9789384568535 (4 Vols. Set)

Typeset by Thomson Press

Printed and bound in India by Thomson Press Ltd.,
New Delhi

Contents

Preface

Integrated pest management is not new to our farmer community. Integrated pest managementt (IPM) has been defined as 'a pest management that in the context of the associated environment and the population dynamics of pest species, utilises all suitable techniques and methods in a compatible manner as possible and maintains pest population at levels below those causing economically unacceptable damage or loss.' Scientific, practical and adoptable measures have been developed and recommended to farmer community. These measures can be used hand in hand with need based and judicious application of pesticides. IPM involves a combination of various measures to ensure effective pest management without disturbing the ecosystem, reducing environmental pollution and eliminate direct and indirect health hazards to human beings.

Plants that are the source of spices became important cash crops over the centuries, and since their introduction, their uses have multiplied. This has resulted in a rise in consumer demand following the popularity for natural flavourings, which has in turn increased interest in their production in temperate and tropical countries. This book is concerned with the pests and diseases and their management through IPM so that profitable production of spice crops at all levels of management, more efficient processing and greater utilization is ensured. Incidence of pests and diseases is very common in spices. The productivity of most of the spice and condiment crops becomes considerably low due to various factors among which infestation by pests and pathogens is a major factor. Several species of insects, mites, nematodes and pathogens attack these crops in the field and also during storage. Synthetic pesticides have been used extensively earlier to reduce crop damage and increase crop profitability leading to pesticide residues in the produce, harmful side effects on beneficial organisms, mortality of pollinators, environmental pollution and resistance to pesticides. Since spices are highvalue and export-oriented commodities, levels of pesticide residues are to be kept well below tolerance limits in view of the stringent standards set by the importing countries. Thus, the growing concern over excessive and indiscriminate pesticide misuse has led to the development of Integrated Pest and Disease Management (IPM) schedules for sustainable management of pests and diseases in recent years. Plant products form an important component in the development of IPM schedules.

A plantation is a large artificially established forest, farm or estate, where crops are grown for sale, often in distant markets rather than for local on-site consumption. Plantation crops belong to many different plant classifications and families. Products from trees include palm oil, citrus fruits, and cocoa; products from bushes include coffee and tea; products from herbs include bananas; and sugar cane is derived from a member of the grass family. The optimal growing conditions, including nutrient and water needs, and threats, including diseases and insects, differ for each crop. They play an important role in view of their export potential as well as domestic requirements and in employment generation and poverty alleviation programmes particularly in rural sector. Though the crops are more sustainable in many ways to growers, processors and to the government, however in the recent years, heavy incidence of pest population is observed causing damage and thereby yield loss. This warrants the farmers to go for pesticide application causing environmental pollution and problems to the ecosystem. The sustainability of plantation crop cultivation is the need of the hour for both farmer as well as processor and finally to the world market. It is essential to manage the pest population to below economic threshold levels to reduce the losses. For this purpose, latest integrated management practices are developed by using the latest low cost technologies which will give sustainability to the plantation industry as a whole.

Though plant products have been evaluated against a many insect and nematode pests of spice and condiment crops they are yet to be experimented for the management of many major pests. Latest advances in the field of IPM in spice and plantation crops are directed to this. There is also an industrial perspective for plant products due to abundance of indigenous resources for plant based pesticides. Education of farmers on proper methods of collection, storage and preparation of plant based preparations would also help to popularize plant products for the management of pests and diseases of spice and condiment crops. The primary aim of pest management should be not to eradicate all organisms, but to manage pests and diseases that may negatively affect production of plantation crops so that they remain at a level that is under an economically and environmentally damaging threshold. Pesticides should be managed to avoid their migration into off-site land or water environments by establishing their use as part of an Integrated Pest Management (IPM) strategy.

Dr. P. Singh

1

Plantation

A plantation is a long, artificially-established forest, farm or estate, where crops are grown for sale, often in distant markets rather than for local on-site consumption. The term *plantation* is informal and not precisely defined. Plantations are grown on a large scale as the crops are grown for commercial purposes, not for local consumption.

Crops grown on plantations include fast-growing trees (often conifers), cotton, coffee, tobacco, sugar cane, sisal, some oil seeds (notably oil palms) and rubber trees. Farms that produce alfalfa, Lespedeza, clover, and other forage crops are usually not called plantations. The term "plantation" has usually not included large orchards (except for banana plantations), but does include the planting of trees for lumber. A plantation is always a monoculture over a large area and does not include extensive naturally occurring stands of plants that have economic value. Because of its large size, a plantation takes advantage of economies of scale. Protectionist policies and natural comparative advantage have contributed to determining where plantations have been located.

Among the earliest examples of plantations were the latifundia of the Roman Empire, which produced large quantities of wine and olive oil for export. Plantation agriculture grew rapidly with the increase in international trade and the development of a worldwide economy that followed the expansion of European colonial empires. Like every economic activity, it has changed over time. Earlier forms of plantation agriculture were associated with large disparities of wealth and income, foreign ownership and political influence, and exploitative social systems such as indentured labour and slavery. The history of the environmental, social and economic issues relating to plantation agriculture are covered in articles that focus on those subjects.

Forestry

Industrial Plantations

Industrial plantations are established to produce a high volume of wood in a short period of time for each society. Plantations are grown by state forestry authorities (for example, the Forestry Commission in Britain) and/or the paper and wood industries and other private landowners (such as Weyerhaeuser, Rayonier and Plum Creek Timber in the United States, Asia Pulp & Paper in Indonesia). Christmas trees are often grown on plantations as well. In southern and southeastern Asia, teak plantations have recently replaced the natural forest.

Industrial plantations are actively managed for the commercial production of forest products. Industrial plantations are usually large-scale. Individual blocks are usually even-aged and often consist of just one or two species. These species can be exotic or indigenous. The plants used for the plantation are often genetically altered for desired traits such as growth and resistance to pests and diseases in general and specific traits, for example in the case of timber species, volumic wood production and stem straightness. Forest genetic resources are the basis for genetic alteration. Selected individuals grown in seed orchards are a good source for seeds to develop adequate planting material.

Wood production on a tree plantation is generally higher than that of natural forests. While forests managed for wood production commonly yield between 1 and 3 cubic metres per hectare per year, plantations of fast-growing species commonly yield between 20 and 30 cubic metres or more per hectare annually; a Grand Fir plantation at Craigvinean in Scotland has a growth rate of 34 cubic metres per hectare per year (Aldhous & Low 1974), and Monterey Pine plantations in southern Australia can yield up to 40 cubic metres per hectare per year (Everard & Fourt 1974). In 2000, while plantations accounted for 5% of global forest, it is estimated that they supplied about 35% of the world's roundwood.

Growth Cycle

- In the first year, the ground is prepared usually by the combination of burning, herbicide spraying, and/or cultivation and then saplings are planted by human crew or by machine. The saplings are usually obtained in bulk from industrial nurseries, which may specialize in selective breeding in order to produce fast growing disease- and pest-resistant strains.

- In the first few years until the canopy closes, the saplings are looked after, and may be dusted or sprayed with fertilizers or pesticides until established.
- After the canopy closes, with the tree crowns touching each other, the plantation is becoming dense and crowded, and tree growth is slowing due to competition. This stage is termed 'pole stage'. When competition becomes too intense (for pine trees, when the live crown is less than a third of the tree's total height), it is time to thin out the section.

  There are several methods for thinning, but where topography permits, the most popular is 'row-thinning', where every third or fourth or fifth row of trees is removed, usually with a harvester. Many trees are removed, leaving regular clear lanes through the section so that the remaining trees have room to expand again. The removed trees are delimbed, forwarded to the forest road, loaded onto trucks, and sent to a mill. A typical pole stage plantation tree is 7–30 cm in diametre at breast height (dbh). Such trees are sometimes not suitable for timber, but are used as pulp for paper and particleboard, and as chips for oriented strand board.
- As the trees grow and become dense and crowded again, the thinning process is repeated. Depending on growth rate and species, trees at this age may be large enough for timber milling; if not, they are again used as pulp and chips.
- Around year 10-60 the plantation is now mature and (in economic terms) is falling off the back side of its growth curve. That is to say, it is passing the point of maximum wood growth per hectare per year, and so is ready for the final harvest. All remaining trees are felled, delimbed, and taken to be processed.
- The ground is cleared, and the cycle is repeated.

Some plantation trees, such as pines and eucalyptus, can be at high risk of fire damage because their leaf oils and resins are flammable to the point of a tree being explosive under some conditions. Conversely, an afflicted plantation can in some cases be cleared of pest species cheaply through the use of a prescribed burn, which kills all lesser plants but does not significantly harm the mature trees.

Criticism of Industrial Plantations

In contrast to a naturally regenerated forest, plantations are typically grown as even-aged monocultures, primarily for timber production.

- Plantations are usually near- or total monocultures. That is, the same species of tree is planted across a given area, whereas a natural forest would contain a far more diverse range of tree species.
- Plantations may include tree species that would not naturally occur in the area. They may include unconventional types such as hybrids, and genetically modified trees may be used sometime in the future. Since the primary interest in plantations is to produce wood or pulp, the types of trees found in plantations are those that are best-suited to industrial applications. For example, pine, spruce and eucalyptus are widely planted far beyond their natural range because of their fast growth rate, tolerance of rich or degraded agricultural land and potential to produce large volumes of raw material for industrial use.
- Plantations are always young forests in ecological terms. Typically, trees grown in plantations are harvested after 10 to 60 years, rarely up to 120 years. This means that the forests produced by plantations do not contain the type of growth, soil or wildlife typical of old-growth natural forest ecosystems. Most conspicuous is the absence of decaying dead wood, a crucial component of natural forest ecosystems.

In the 1970s, Brazil began to establish high-yield, intensively managed, short rotation plantations. These types of plantations are sometimes called fast-wood plantations or fibre farms and often managed on a short-rotation basis, as little as 5 to 15 years. They are becoming more widespread in South America, Asia and other areas. The environmental and social impacts of this type of plantation has caused them to become controversial. In Indonesia, for example, large multi-national pulp companies have harvested large areas of natural forest without regard for regeneration. From 1980 to 2000, about 50% of the 1.4 million hectares of pulpwood plantations in Indonesia have been established on what was formerly natural forest land.

The replacement of natural forest with tree plantations has also caused social problems. In some countries, again, notably Indonesia, conversions of natural forest are made with little regard for rights of the local people. Plantations established purely for the production of fibre provide a much narrower range of services than the original natural forest for the local people. India has sought to limit this damage by limiting the amount of land owned by one entity and, as a result, smaller plantations are owned by local farmers who then sell the wood to larger companies. Some large environmental organizations are critical of these high-yield plantations and are running an anti-plantation campaign, notably the Rainforest Action Network and Greenpeace.

Farm or Home Plantations

Farm or home plantations are typically established for the production of timber and fire wood for home use and sometimes for sale. Management may be less intensive than with Industrial plantations. In time, this type of plantation can become difficult to distinguish from naturally regenerated forest.

Teak and bamboo plantations in India have given good results and an alternative crop solution to farmers of central India, where conventional farming was popular. But due to rising input costs of farming many farmers have done teak and bamboo plantations which require very little water (only during first two years). Teak and bamboo have legal protection from theft. Bamboo, once planted, gives output for 50 years till flowering occurs. Teak requires 20 years to grow to full maturity and fetch returns.

Environmental Plantations

These may be established for watershed or soil protection. They are established for erosion control, landslide stabilization and windbreaks. Such plantations are established to foster native species and promote forest regeneration on degraded lands as a tool of environmental restoration.

Ecological Impact

Probably the single most important factor a plantation has on the local environment is the site where the plantation is established. If natural forest is cleared for a planted forest then a reduction in biodiversity and loss of habitat will likely result. In some cases, their establishment may involve draining wetlands to replace mixed hardwoods that formerly predominated with pine species. If a plantation is established on abandoned agricultural land, or highly degraded land, it can result in an increase in both habitat and biodiversity. A planted forest can be profitably established on lands that will not support agriculture or suffer from lack of natural regeneration.

The tree species used in a plantation is also an important factor. Where non-native varieties or species are grown, few of the native fauna are adapted to exploit these and further biodiversity loss occurs. However, even non-native tree species may serve as corridors for wildlife and act as a buffer for native forest, reducing edge effect.

Once a plantation is established, how it is managed becomes the important environmental factor. The single most important factor of management is the rotation period. Plantations harvested on longer rotation periods (30 years or more) can provide similar benefits to a naturally regenerated forest managed for wood production, on a similar rotation. This is especially true if native species are used. In the case of exotic species, the habitat can be improved significantly if the impact is mitigated by measures such as leaving blocks of native species in the plantation, or retaining corridors of natural forest. In Brazil, similar measures are required by government regulations.

Plantations and Natural Forest Loss

Many forestry experts claim that the establishment of plantations will reduce or eliminate the need to exploit natural forest for wood production. In principle this is true because due to the high productivity of plantations less land is needed. Many point to the example of New Zealand, where 19% of the forest area provides 99% of the supply of industrial round wood. It has been estimated that the world's demand for fibre could be met by just 5% of the world forest (Sedjo & Botkin 1997). However in practice, plantations are replacing natural forest, for example in Indonesia.

According to the FAO, about 7% of the natural closed forest being lost in the tropics is land being converted to plantations. The remaining 93% of the loss is land being converted to agriculture and other uses. Worldwide, an estimated 15% of plantations in tropical countries are established on closed canopy natural forest.

In the Kyoto Protocol, there are proposals encouraging the use of plantations to reduce carbon dioxide levels (though this idea is being challenged by some groups on the grounds that the sequestered CO_2 is eventually released after harvest).

Other Types of Plantation

Crops may be called plantation crops because of their association with a specific type of farming economy. Most of these involve a large landowner, raising crops with economic value rather than for subsistence, with a number of employees carrying out the work. Often it referred to crops newly introduced to a region. In past times it has been associated with slavery, indentured labour, and other economic models of high inequity. However, arable and dairy farming are both usually (but not always) excluded from such definitions. A comparable economic structure in antiquity was the *latifundia* that produced commercial quantities of olive oil or wine, for export. One plantation crop is bananas and there are others as well.

High Value Food Crops

Plantings of a number of trees or shrubs grown for food or beverage, including tea, coffee, and cacao are generally called plantations. Some spice and high value crops grown from permanent perennial stock, such as black pepper, may also be so called. When the holding belongs to a single individual, that person may be called a *planter*.

Sugar: Sugar plantations were highly valued in the Caribbean by the British and French colonists in the 17th and 18th centuries and the use of sugar in Europe rose during this period. Sugarcane is still an important crop in Cuba. Sugar plantations also arose in countries such as Barbados and Cuba because of the natural endowments that they had. These natural endowments included soil that was conducive to growing sugar and a high marginal product of labour realized through the increasing number of slaves.

Rubber: Plantings of para rubber, the tree *Hevea brasiliensis*, are usually called plantations.

Oil Palm: Oil palm agriculture is rapidly expanding across wet tropical regions, and is usually developed at plantation scale.

Orchards: Fruit orchards are sometimes considered to be plantations.

Arable Crops: These include tobacco, sugarcane, pineapple, and cotton, especially in historical usage.

Before the rise of cotton in the American South, indigo and rice were also sometimes called plantation crops.

Fishing Plantations in Newfoundland and Labrador

When Newfoundland was colonized by England in 1610, the original colonists were called "Planters" and their fishing rooms were known as "fishing plantations". These terms were used well into the 20th century.

The following three plantations are maintained by the Government of Newfoundland and Labrador as provincial heritage sites:

- Sea-Forest Plantation was a 17th-century fishing plantation established at Cuper's Cove (present-day Cupids) under a royal charter issued by King James I.

- Mockbeggar Plantation is an 18th-century fishing plantation at Bonavista.
- Pool Plantation a 17th-century fishing plantation maintained by Sir David Kirke and his heirs at Ferryland. The plantation was destroyed by French invaders in 1696.

Other fishing plantations:

- Bristol's Hope Plantation, a 17th-century fishing plantation established at Harbour Grace, created by the Bristol Society of Merchant-Adventurers.
- Benger Plantation, an 18th-century fishing plantation maintained by James Benger and his heirs at Ferryland. It was built on the site of Georgia plantation.
- Piggeon's Plantation, an 18th-century fishing plantation maintained by Ellias Piggeon at Ferryland.

Plantation Economy

A plantation economy is an economy based on agricultural mass production, usually of a few staple products grown on large farms called plantations. Plantation economies rely on the export of cash crops as a source of income. Prominent plantation crops included cotton, rubber, sugar cane, tobacco, figs, rice, kapok, sisal and species in the genus *Indigofera*, used to produce indigo dye. The longer a crop's harvest period, the more efficient plantations become. Economies of scale are also achieved when the distance to market is long and when the crop's size is reduced. Plantation crops also differ in that they usually need processing immediately after harvesting. Sugar, tea sisal and palm oil are most suited to plantations, while coconuts, rubber and cotton are suitable to a lesser extent.

Slavery, Para-slavery and Plantations

Slave labour extracted from forcibly transported Africans was used extensively to work on early plantations (such as cotton and sugar plantations) in the American colonies and the United States, throughout the Caribbean, the Americas and in European-occupied areas of Africa. Several notable historians and economists such as Eric Williams, Walter Rodney and Karl Marx contend that the global capitalist economy was largely founded on the creation and produce of thousands of slave labour camps based in colonial plantations exploiting tens of millions of abducted Africans.

In modern times, the low wages typically paid to plantation workers are the basis of plantation profitability in some areas. Sugarcane plantations in the Caribbean and Brazil, worked by slave labour, were also examples of the plantation system.

In more recent times, overt slavery has been replaced by "para-slavery" or slavery-in-kind, including the sharecropping system. At its most extreme, workers are in "debt bondage": they must work to pay off a debt at such punitive interest rates that it may never be paid off. Others work unreasonably long hours and are paid subsistence wages that (in practice) may only be spent in the company store.

In Brazil, a sugarcane plantation was termed an *engenho* ("engine"), and the 17th-century English usage for organized colonial production was "factory". Such colonial social and economic structures are discussed at Plantation economy. Sugar workers on plantations in Cuba and elsewhere in the Caribbean lived in company towns known as Bateys.

Plantations in the Antebellum American South

In the American South, antebellum plantations were centred on a "plantation house", the residence of the owner, where important business was conducted. Slavery and plantations had different characteristics in different regions of the South. As the Upper South of the Chesapeake Bay Colony developed first, historians of the antebellum South defined planters as those who held 20 or more slaves. Major planters held many more, especially in the Deep South as it developed. The majority of slaveholders held 10 or fewer slaves, often just a few to labour domestically. By the late 18th century, most planters in the Upper South had switched from exclusive tobacco cultivation to mixed crop production, both because tobacco had exhausted the soil and because of changing markets. The shift away from tobacco meant they had slaves in excess of the number needed for labour, and they began to sell them in the internal slave trade.

There was a variety of domestic architecture on plantations. The largest and wealthiest planter families, for instance, those with estates fronting on the James River in Virginia, constructed mansions in brick and Georgian style, e.g. Shirley Plantation. Common or smaller planters in the late 18th and 19th century had more modest wood frame buildings, such as Southall Plantation in Charles City County.

In the Low Country of South Carolina, by contrast, even before the American Revolution, planters holding large rice and cotton plantations in South Carolina typically owned hundreds of slaves. In Charleston and Savannah, the elite also held numerous slaves to work as household servants. The 19th-century development of the Deep South for cotton cultivation depended on large plantations with much more acreage than was typical of the Chesapeake Bay area, and for labour, planters held hundreds of slaves.

Until December 1865 slavery was legal in parts of the United States. Most slaves were employed in agriculture, and "planter" was a term commonly used to describe a farmer with many slaves.

The term "planter" has no universally accepted definition but academic historians have defined it to identify the elite class, "a landowning farmer of substantial means." In the "Black Belt" counties of Alabama and Mississippi, the terms ""planter" and "farmer" were often synonymous. Robert Fogel and Stanley Engerman define large planters as owning over 50 slaves, and medium planters as owning between 16 and 50 slaves. In his study of Black Belt counties in Alabama, Jonathan Wiener defines planters by ownership of real property, rather than of slaves.

A planter, for Wiener, owned at least $10,000 worth of real estate in 1850 and $32,000 worth in 1860, equivalent to about the top 8 percent of landowners. In his study of southwest Georgia, Lee Formwalt also defines planters in size of land holdings rather than slaves. Formwalt's planters are in the top 4.5 percent of landowners, translating into real estate worth $6,000 or more in 1850, $24,000 or more in 1860, and $11,000 or more in 1870. In his study of Harrison County, Texas, Randolph B. Campbell classifies large planters as owners of 20 slaves, and small planters as owners of between ten and 19 slaves. In Chicot and Phillips Counties, Arkansas, Carl H. Moneyhon defines large planters as owners of twenty or more slaves, and six hundred or more acres.

Farm or Home Plantations

A tree farm is a privately owned forest managed for timber production. The term, tree farm, also is used to refer to tree plantations, tree nurseries, and Christmas tree farms.

American Tree Farm System

The American Tree Farm System (ATFS) is the largest and oldest woodland certification system in America. It is internationally recognized by the Programme for the Endorsement of Forest Certification and meets strict third-party certification standards. It is one of three certification systems currently recognized in the United States (the others include the Forest Stewardship Council and the Sustainable Forestry Initiative). ATFS specializes in certifying private forests, primarily those held by individuals and families and currently certifies over 26 million acres (110,000 km²) of forestland. The ATFS Standard for Certification is owned by the American Forest Foundation, a national nonprofit organization focused on environmental education and promoting sustainable stewardship of America's woodlands.

***Figure:** A well-managed ATFS-certified tree farm in Virginia provides a variety of habitats for wildlife while sustainably producing wood*

History

The American Tree Farm System was established in 1942 in an effort to promote resources on private land, ensuring plentiful fibre production for timber and paper companies. With declining virgin saw timber available, the industry began to promote forestry practices to ensure sufficient fibre production for the future. Prior to 1941, the majority of fibre came from industrial lands.

The first tract of land labelled as a Tree Farm was organized and marketed by the Weyerhaeuser Company to help change public attitudes toward timber production and protect natural resources from forest fires and other natural disasters. The title of "tree farm" was chosen in large part because Weyerhaeuser felt that the 1940s public understood farming as crop production, and similarly tree farming was focused on producing more timber, with frequent replanting post-harvest.

The early sponsors of the tree-farming movement defined it as "privately owned forest-land dedicated to the growing of forest crops for commercial purposes, protected and managed for continuous production of forest products."

In the early 1940s the concept of "tree-farming" on private land was promoted by the National Lumber Manufacturers Association in an organized campaign to engage timberland owners in conservative timber production.

Throughout its history, ATFS has relied on celebrity Tree Farmers to relay its message to the public. Celebrities include actor Andy Griffith, actress Andie MacDowell, former President Jimmy Carter, and Rolling Stone keyboardist Chuck Leavell.

Current

Since 1941, the system has shifted to focus on whole stewardship, rather than strictly fibre production. According to the Standards of Certification for ATFS, woodland owners must own 10 or more acres and have a management plan. In that management plan, woodland owners must recognize wildlife habitat, protection of water quality, threatened and endangered species, and sustainable harvest levels. The certification standard is subject to multi-stakeholder involvement in the development and revision of the standard, third-party audits, and a publicly available certification of audit summaries.

It should be noted that the minimum acreage to qualify for a tree farm refers to "woodland" i.e., forested land. So acreage which includes grazing or other non-wooded lands must have at least 10 acres in forest to qualify. Furthermore, programmes in different areas which support tree farming activities may require larger forested acreages as well as additional criteria. For example, The Forest Ag Programme in Colorado requires the following standards:

To be eligible for the Forest Ag Programme, properties must meet several criteria:

- The landowner must perform forest management activities to produce tangible wood products for the primary purpose of obtaining a monetary profit. Tangible wood products include transplants, Christmas trees and boughs, sawlogs, posts, poles and firewood.
- The landowner must have at least 40 forested acres.
- The landowner must submit a Colorado State Forest Service-approved forest management plan that is prepared by a professional forester or natural resources professional.

Landowners must annually submit (1) a request for inspection, (2) an inspection fee, (3) an accomplishment report, and (4) an annual work plan for the following year, and have the enrolled property inspected by a CSFS forester.

As a programme of the American Forest Foundation (AFF), the American Tree Farm System focuses on the long-term sustainability of America's forests in ecological and economic terms. The vision statement of AFF states, "AFF is committed to creating a future where North American forests are sustained by the public that understand and values the social, economic, and environmental benefits they provide to our communities, our nation, and the world."

The network of over 90,000 woodland owners is organized through state committees and governed at the national level. Currently 45 of the 50 states have committees. Alaska, Arizona, Hawaii, North Dakota and Utah currently do not have programmes. With national coordination, ATFS strives to "work on-the-ground with families...to promote stewardship and protect our nation's forest heritage." The state networks also include tree farm inspectors, who certify the forests and conduct outreach efforts on behalf of ATFS and partnered organizations.

Each year ATFS hosts a National Tree Farmer Convention and awards an individual or family with the National Outstanding Tree Farmer of the Year award. It also awards a National Outstanding Inspector Award to a resource professional who has demonstrated exceptional outreach efforts to engage landowners and the general public in sustainable forestry.

Tree Farming and Climate Change

A forest sequesters carbon in its trees. The forest removes carbon dioxide from the air as trees grow and returns it to the air as trees die and rot or burn. As long as the forest is experiencing net growth, the forest is reducing the amount of carbon dioxide, the leading greenhouse gas, from the air. Furthermore, if timber is regularly removed from the forest and turned into lasting wood products, those products continue sequestering carbon, while the replacement tree farm trees absorb more carbon dioxide, thus effecting a continuous reduction in greenhouse gas.

Because tree farms are managed to enhance rapid growth, a tree farm tends to sequester carbon more quickly than an unmanaged forest, considering only the sequestration side of the equation and not the carbon release due to rot, fire, or harvest. The fact that managed woodlands tend to be younger and younger trees grow faster and die less contributes to this distinction.

While tree farms absorb large amounts of CO_2, the long-term sequestration of this carbon depends on what is done with the harvested materials. Forests continue to absorb atmospheric carbon for centuries if left undisturbed.

The USDA has an online calculator for how much carbon is sequestered in various types of forests.

CO2 and Forest Health

Carbon dioxide is a primary building material for plant tissue and is required to make plants grow fast and strong, so presumably higher levels of CO_2 in the air as a result of burning fossil fuels would make forests grow faster. Duke University did a study where they dosed a loblolly pine plantation with elevated levels of CO2. The studies showed that the pines did indeed grow faster and stronger. They were also less prone to damage during ice storms, which is a factor that limits loblolly growth farther north.

The forest did relatively better during dry years. The hypothesis is that the limiting factors in the growth of the pines are nutrients such as nitrogen, which is in deficit on much of the pine land in the Southeast. In dry years, however, the trees don't bump up against those factors since they are growing more slowly because water is the limiting factor. When rain is plentiful trees reach the limits of the site's nutrients and the extra CO2 isn't beneficial. Most forest soils in Southeastern region are deficient in nitrogen and phosphorus as well as trace minerals. Pine forests often sit on land that was used for cotton, corn or tobacco. Since these crops depleted originally shallow and infertile soils, tree farmers must work to improve soils. In addition to better fertilization, biosolids present an innovative solution. Biosolids are treated sewage from municipal or agricultural sources such as chicken and hog operations in Virginia and North Carolina. Though biosolids have the potential to improve soils and lead to improved tree growth barriers to adoption include regulation and inertia.

Teak

Teak is a tropical hardwood tree species *Tectona grandis*. The species is placed in the family Lamiaceae. *Tectona grandis* is a large, deciduous tree that is dominant in mixed hardwood forests. It has small, fragrant white flowers and papery leaves that are often hairy on the lower surface. It is sometimes known as the "Burmese Teak". Teak wood has a leather-like smell when it is freshly milled.

Teak timber is particularly valued for its durability and water resistance, and is used for boat building, exterior construction, veneer, furniture, carving, turnings, and other small wood projects. *Tectona grandis* is native to south and southeast Asia, mainly India, Indonesia, Malaysia, Thailand and Burma, but is naturalized and cultivated in many countries in Africa and the Caribbean. Burma accounts for nearly one third of the world's total teak production.

Description

Teak is a large, deciduous tree up to 40 m (131 ft) tall with gray to grayish brown branches. Leaves are ovate-elliptic to ovate, 15–45 cm (5.9–17.7 in) long by 8–23 cm (3.1–9.1 in) wide, and are held on robust petioles that are 2–4 cm (0.8–1.6 in) long. Leaf margins are entire.

Fragrant white flowers are borne on 25–40 cm (10–16 in) long by 30 cm (12 in) wide panicles from June to August. The corolla tube is 2.5–3 mm long with 2 mm wide obtuse lobes. *Tectona grandis* sets fruit from September to December; fruits are globose and 1.2-1.8 cm in diametre. Flowers are weakly protandrous in that the anthers precede the stigma in maturity and pollen is shed within a few hours of the flower opening.

The flowers are primarily entomophilous (insect-pollinated), but can occasionally be anemophilous (wind-pollinated). A 1996 study found that in its native range in Thailand, the major pollinator were species in the *Ceratina* genus of bees.

Distribution and Habitat

Tectona grandis is one of three species in the genus *Tectona*. The other two species, *T. hamiltoniana* and *T. philippinensis*, are endemics with relatively small native distributions in Myanmar and the Philippines, respectively. *Tectona grandis* is native to India, Indonesia, Myanmar, northern Thailand, and northwestern Laos.

Tectona grandis is found in a variety of habitats and climatic conditions from arid areas with only 500 mm of rain per year to very moist forests with up to 5,000 mm of rain per year. Typically, though, the annual rainfall in areas where teak grows averages 1,250-1,650 mm with a 3-5 month dry season.

Botanical History

Tectona grandis was first formally described by Carl Linnaeus the Younger in his 1782 work *Supplementum Plantarum*. In 1975, Harold Norman Moldenke published new descriptions of four forms of this species in the journal *Phytologia*. Moldenke described each form as varying slightly from the type specimen: *T. grandis* f. *canescens* is distinguished from the type material by being densely canescent, or covered in hairs, on the underside of the leaf, *T. grandis* f. *pilosula* is distinct from the type material in the varying morphology of the leaf veins, *T. grandis* f. *punctata* is only hairy on the larger veins on the underside of the leaf, and *T. grandis* f. *tomentella* is noted for its dense yellowish tomentose hairs on the lower surface of the leaf.

Cultivation

Teak's natural oils make it useful in exposed locations, and make the timber termite and pest resistant. Teak is durable even when not treated with oil or varnish. Timber cut from old teak trees was once believed to be more durable and harder than plantation grown teak. Studies have shown that plantation teak performs on par with old-growth teak in erosion rate, dimensional stability, warping, and surface checking, but is more susceptible to colour change from UV exposure.

The vast majority of commercially harvested teak is grown on teak plantations found in Indonesia and controlled by Perum Perhutani (a state owned forest enterprise) that manages the country's forests. The primary use of teak harvested in Indonesia is in the production of outdoor teak furniture for export. Nilambur in Kerala, India is also a major producer of Teak of fine quality, holds the world's oldest Teak plantation.

Teak consumption raises a number of environmental concerns, such as the disappearance of rare old-growth teak. However, its popularity has led to growth in sustainable plantation teak production throughout the seasonally dry tropics in forestry plantations. The Forest Stewardship Council offers certification of sustainably grown and harvested teak products. Propagation of teak via tissue culture for plantation purposes is commercially viable.

Teak plantations were widely established in Equatorial Africa during the Colonial era. These timber resources, as well as the oil reserves, are at the heart of the current (2014) South Sudanese conflict.

Much of the world's teak is exported by Indonesia and Myanmar. There is also a rapidly growing plantation grown market in Central America (Costa Rica) and South America.

Hyblaea puera, a moth native to southeast Asia, is a teak pest whose caterpillar feeds on teak and other species of trees common in the region.

Uses

Teak is a yellowish brown timber with good grain and texture. Teak, though easily worked, can cause severe blunting on edged tools because of the presence of silica in the wood. Teak is often an effective material for the construction of both indoor and outdoor furniture. Teak's high oil content, high tensile strength and tight grain makes it particularly suitable for outdoor furniture applications. Over time teak can mature to a silvery-grey finish, especially when exposed to sunlight. It is used in the manufacture of outdoor furniture, boat decks, and other articles where weather resistance is desired. It is also used for cutting boards, indoor flooring, countertops and as a veneer for indoor furnishings. Teak is used extensively in India to make doors and window frames, furniture, and columns and beams in old type houses. It is very resistant to termite attacks. Mature teak fetches a very good price. It is grown extensively by forest departments of different states in forest areas.

Leaves of the teak wood tree are used in making Pellakai gatti (jackfruit dumpling), where batter is poured into a teak leaf and is steamed. This type of usage is found in the coastal district of Udupi in the Tulunadu region in South India. The leaves are also used in gudeg, a dish of young jackfruit made in Central Java, Indonesia, and give the dish its dark brown colour.

Teak is used as a food plant by the larvae of moths of the genus *Endoclita* including *E. aroura*, *E. chalybeatus*, *E. damor*, *E. gmelina*, *E. malabaricus*, *E. sericeus* and *E. signifer* and other Lepidoptera including Turnip Moth.

Uses in Boatbuilding

Teak has been used as a boatbuilding material for over 150 years. In addition to relatively high strength, teak is also highly resistant to rot, fungi and mildew. In addition, teak has a relatively low shrinkage ratio, which makes it excellent for applications where it undergoes periodic changes in moisture. Teak has the unusual properties of being both an excellent structural timber for framing, planking, etc., while at the same time being easily worked, unlike some other similar woods such as purpleheart, and finished to a high degree. For this reason, it is also prized for the trim work on boat interiors. Due to the oily nature of the wood, care must be taken to properly prepare the wood before gluing.

When used on boats, teak is also very flexible in the finishes that may be applied. One option is to use no finish at all, in which case the wood will naturally weather to a pleasing silver-grey. The wood may also be oiled with a finishing agent such as linseed or tung oil. This results in a pleasant, somewhat bland finish. Finally, teak may also be varnished for a deep, lustrous glow.

Teak is also used extensively in boat decks, as it is extremely durable and requires very little maintenance. The teak tends to wear in to the softer 'summer' growth bands first, forming a natural 'non-slip' surface. Any sanding is therefore only damaging. Use of modern cleaning compounds, oils or preservatives will shorten the life of the teak, as it contains natural teak-oil a very small distance below the white surface. Wooden boat experts will only wash the teak with salt water, and re-caulk when needed. This cleans the deck, and prevents it from drying out and the wood shrinking. The salt helps it absorb and retain moisture, and prevents any mildew and algal growth. People with poor knowledge often over-maintain the teak, and drastically shorten its life.

Alternatives to Teak

Due to the increasing cost of teak, various alternatives have been employed. These include purpleheart, iroko, and angelique.

Propagation

Figure: *Tree in new leaves in Kolkata, West Bengal, India.*

Teak is propagated mainly from seeds. Germination of the seeds involves pretreatment to remove dormancy arising from the thick pericarp. Pretreatment involves alternate wetting and drying of the seed. The seeds are soaked in water for 12 hours and then spread to dry in the sun for 12 hours. This is repeated for 10–14 days and then the seeds are sown in shallow germination beds of coarse peat covered by sand. The seeds then germinate after 15 to 30 days.

Clonal propagation of teak has been successfully done through grafting, rooted stem cuttings and micro propagation. While bud grafting on to seedling root stock has been the method used for establishing clonal seed orchards that enables assemblage of clones of the superior trees to encourage crossing, rooted stem cuttings and micro propagated plants are being increasingly used around the world for raising clonal plantations.

Bamboo

Bamboo is a tribe of flowering perennial evergreen plants in the grass family Poaceae, subfamily Bambusoideae, tribe Bambuseae. Giant bamboos are the largest members of the grass family.

In bamboos, the internodal regions of the stem are hollow and the vascular bundles in the cross section are scattered throughout the stem instead of in a cylindrical arrangement.

The dicotyledonous woody xylem is also absent. The absence of secondary growth wood causes the stems of monocots, even of palms and large bamboos, to be columnar rather than tapering.

Bamboos are some of the fastest-growing plants in the world, due to a unique rhizome-dependent system. Bamboos are of notable economic and cultural significance in South Asia, Southeast Asia and East Asia, being used for building materials, as a food source, and as a versatile raw product. Bamboo as a higher compressive strength than wood, brick or concrete (on an ounce for ounce basis) and has a tensile strength that rivals steel.

Genus and Geography

More than 10 genera are divided into about 1,450 species. Bamboo species are found in diverse climates, from cold mountains to hot tropical regions. They occur across East Asia, from 50°N latitude in Sakhalin through to Northern Australia, and west to India and the Himalayas. They also occur in sub-Saharan Africa, and in the Americas from the mid-Atlantic United States south to Argentina and Chile, reaching their southernmost point at 47°S latitude. Continental Europe is not known to have any native species of bamboo.

Recently, some attempts have been made to grow bamboo on a commercial basis in the Great Lakes region of east-central Africa, especially in Rwanda. Companies in the United States are growing, harvesting and distributing species such as Henon and Moso. Bamboo grows in two main forms: the woody bamboos (Arundinarieae and Bambuseae) and the understory herbaceous bamboos (Olyreae).

Molecular analysis suggests that there are 3–5 major lineages of bamboo. Four major lineages are currently recognized: temperate woody, paleotropical woody, neotropical woody and herbaceous.

Ecology

Bamboo is one of the fastest-growing plants on Earth, with reported growth rates of 250 cm (98 in) in 24 hours. However, the growth rate is dependent on local soil and climatic conditions, as well as species, and a more typical growth rate for many commonly cultivated bamboos in temperate climates is in the range of 3–10 centimetres (1.2–3.9 in) per day during the growing period.

Primarily growing in regions of warmer climates during the late Cretaceous period, vast fields existed in what is now Asia. Some of the largest timber bamboo can grow over 30 m (98 ft) tall, and be as large as 15–20 cm (5.9–7.9 in) in diametre. However, the size range for mature bamboo is species dependent, with the smallest bamboos reaching only several inches high at maturity. A typical height range that would cover many of the common bamboos grown in the United States is 4.6–12 metres (15–39 ft), depending on species. Anji County of China, known as the "Town of Bamboo", provides the optimal climate and soil conditions to grow, harvest, and process some of the most valued bamboo poles available worldwide.

Unlike all trees, individual bamboo stems, or culms, emerge from the ground at their full diametre and grow to their full height in a single growing season of three to four months. During these several months, each new shoot grows vertically into a culm with no branching out until the majority of the mature height is reached. Then, the branches extend from the nodes and leafing out occurs. In the next year, the pulpy wall of each culm slowly hardens. During the third year, the culm hardens further. The shoot is now considered a fully mature culm.

Over the next 2–5 years (depending on species), fungus begins to form on the outside of the culm, which eventually penetrates and overcomes the culm. Around 5–8 years later (species and climate dependent), the fungal growths cause the culm to collapse and decay. This brief life means culms are ready for harvest and suitable for use in construction within about three to seven years. I

ndividual bamboo culms do not get any taller or larger in diametre in subsequent years than they do in their first year, and they do not replace any growth lost from pruning or natural breakage. Bamboos have a wide range of hardiness depending on species and locale. Small or young specimens of an individual species will produce small culms initially. As the clump and its rhizome system mature, taller and larger culms will be produced each year until the plant approaches its particular species limits of height and diametre.

Many tropical bamboo species will die at or near freezing temperatures, while some of the hardier or so-called temperate bamboos can survive temperatures as low as “29 °C (“20 °F). Some of the hardiest bamboo species can be grown in places as cold as USDA Plant Hardiness Zones 5–6, although they typically will defoliate and may even lose all above-ground growth, yet the rhizomes will survive and send up shoots again the next spring. In milder climates, such as USDA Zone 8 and above, some hardy bamboo may remain fully leafed out year-round.

Mass Flowering

Most bamboo species flower infrequently. In fact, many bamboos only flower at intervals as long as 65 or 120 years. These taxa exhibit mass flowering (or gregarious flowering), with all plants in a particular species flowering worldwide over a several-year period. The longest mass flowering interval known is 130 years, and it is for the species *Phyllostachys bambusoides* (Sieb. & Zucc.). In this species, all plants of the same stock flower at the same time, regardless of differences in geographic locations or climatic conditions, and then the bamboo dies. The lack of environmental impact on the time of flowering indicates the presence of some sort of “alarm clock” in each cell of the plant which signals the diversion of all energy to flower production and the cessation of vegetative growth. This mechanism, as well as the evolutionary cause behind it, is still largely a mystery.

One hypothesis to explain the evolution of this semelparous mass flowering is the *predator satiation hypothesis* which argues that by fruiting at the same time, a population increases the survival rate of their seeds by flooding the area with fruit, so, even if predators eat their fill, seeds will still be left over. By having a flowering cycle longer than the lifespan of the rodent predators, bamboos can regulate animal populations by causing starvation during the period between flowering events. Thus the death of the adult clone is due to resource exhaustion, as it would be more effective for parent plants to devote all resources to creating a large seed crop than to hold back energy for their own regeneration.

Another, the *fire cycle hypothesis*, argues that periodic flowering followed by death of the adult plants has evolved as a mechanism to create disturbance in the habitat, thus providing the seedlings with a gap in which to grow. This argues that the dead culms create a large fuel load, and also a large target for lightning strikes, increasing the likelihood of wildfire. Because bamboos can be aggressive as early successional plants, the seedlings would be able to outstrip other plants and take over the space left by their parents.

However, both have been disputed for different reasons. The predator satiation hypothesis does not explain why the flowering cycle is 10 times longer than the lifespan of the local rodents, something not predicted. The bamboo fire cycle hypothesis is considered by a few scientists to be unreasonable; they argue that fires only result from humans and there is no natural fire in India. This notion is considered wrong based on distribution of lightning strike data during the dry season throughout India. However, another argument against this is the lack of precedent for any living organism to harness something as unpredictable as lightning strikes to increase its chance of survival as part of natural evolutionary progress.

The mass fruiting also has direct economic and ecological consequences, however. The huge increase in available fruit in the forests often causes a boom in rodent populations, leading to increases in disease and famine in nearby human populations. For example, devastating consequences occur when the *Melocanna bambusoides* population flowers and fruits once every 30–35 years around the Bay of Bengal. The death of the bamboo plants following their fruiting means the local people lose their building material, and the large increase in bamboo fruit leads to a rapid increase in rodent populations. As the number of rodents increases, they consume all available food, including grain fields and stored food, sometimes leading to famine. These rats can also carry dangerous diseases, such as typhus, typhoid, and bubonic plague, which can reach epidemic proportions as the rodents increase in number. The relationship between rat populations and bamboo flowering was examined in a 2009 Nova documentary Rat Attack.

In any case, flowering produces masses of seeds, typically suspended from the ends of the branches. These seeds will give rise to a new generation of plants that may be identical in appearance to those that preceded the flowering, or they may produce new cultivars with different characteristics, such as the presence or absence of striping or other changes in coloration of the culms.

As Animal Diet

Soft bamboo shoots, stems, and leaves are the major food source of the giant panda of China, the red panda of Nepal and the bamboo lemurs of Madagascar. Rats will eat the fruits as described above. Mountain gorillas of Africa also feed on bamboo, and have been documented consuming bamboo sap which was fermented and alcoholic; chimps and elephants of the region also eat the stalks.

The larvae of the bamboo borer (the moth *Omphisa fuscidentalis*) of Laos, Myanmar, Thailand and Yunnan Province, China, feeds off the pulp of live bamboo. In turn, these caterpillars are considered a local delicacy.

Cultivation

Commercial Timber: Timber is harvested from both cultivated and wild stands, and some of the larger bamboos, particularly species in the genus *Phyllostachys*, are known as "timber bamboos".

Harvesting

Bamboo used for construction purposes must be harvested when the culms reach their greatest strength and when sugar levels in the sap are at their lowest, as high sugar content increases the ease and rate of pest infestation.

Harvesting of bamboo is typically undertaken according to the following cycles:

1) Life cycle of the culm: As each individual culm goes through a 5– to 7-year life cycle, culms are ideally allowed to reach this level of maturity prior to full capacity harvesting. The clearing out or thinning of culms, particularly older decaying culms, helps to ensure adequate light and resources for new growth. Well-maintained clumps may have a productivity three to four times that of an unharvested wild clump.
2) Life cycle of the culm: Consistent with the life cycle described above, bamboo is harvested from two to three years through to five to seven years, depending on the species.
3) Annual cycle: As all growth of new bamboo occurs during the wet season, disturbing the clump during this phase will potentially damage the upcoming crop. Also during this high rainfall period, sap levels are at their highest, and then diminish towards the dry season. Picking immediately prior to the wet/growth season may also damage new shoots. Hence, harvesting is best at the end of the dry season, a few months prior to the start of the wet.
4) Daily cycle: During the height of the day, photosynthesis is at its peak, producing the highest levels of sugar in sap, making this the least ideal time of day to harvest. Many traditional practitioners believe the best time to harvest is at dawn or dusk on a waning moon.

Leaching

Leaching is the removal of sap after harvest. In many areas of the world, the sap levels in harvested bamboo are reduced either through leaching or postharvest photosynthesis. Examples of this practice include:

1. Cut bamboo is raised clear of the ground and leaned against the rest of the clump for one to two weeks until leaves turn yellow to allow full consumption of sugars by the plant.
2. A similar method is undertaken, but with the base of the culm standing in fresh water, either in a large drum or stream to leach out sap.
3. Cut culms are immersed in a running stream and weighted down for three to four weeks.
4. Water is pumped through the freshly cut culms, forcing out the sap (this method is often used in conjunction with the injection of some form of treatment).

In the process of water leaching, the bamboo is dried slowly and evenly in the shade to avoid cracking in the outer skin of the bamboo, thereby reducing opportunities for pest infestation.

Durability of bamboo in construction is directly related to how well it is handled from the moment of planting through harvesting, transportation, storage, design, construction and maintenance. Bamboo harvested at the correct time of year and then exposed to ground contact or rain, will break down just as quickly as incorrectly harvested material.

Ornamental Bamboos

The two general patterns for the growth of bamboo are "clumping" (sympodial) and "running" (monopodial). Clumping bamboo species tend to spread slowly, as the growth pattern of the rhizomes is to simply expand the root mass gradually, similar to ornamental grasses. "Running" bamboos, on the other hand, need to be controlled during cultivation because of their potential for aggressive behaviour. They spread mainly through their roots and/or rhizomes, which can spread widely underground and send up new culms to break through the surface. Running bamboo species are highly variable in their tendency to spread; this is related to both the species and the soil and climate conditions. Some can send out runners of several metres a year, while others can stay in the same general area for long periods. If neglected, over time they can cause problems by moving into adjacent areas.

Bamboos seldom and unpredictably flower, and the frequency of flowering varies greatly from species to species. Once flowering takes place, a plant will decline and often die entirely. Although there are always a few species of bamboo in flower at any given time, collectors desiring to grow specific bamboo typically obtain their plants as divisions of already-growing plants, rather than waiting for seeds to be produced.

Regular observations will indicate major growth directions and locations. Once the rhizomes are cut, they are typically removed; however, rhizomes take a number of months to mature, and an immature, severed rhizome will usually cease growing if left in-ground. If any bamboo shoots come up outside of the bamboo area afterwards, their presence indicates the precise location of the removed rhizome. The fibrous roots that radiate from the rhizomes do not produce more bamboo.

Bamboo growth is also controlled by surrounding the plant or grove with a physical barrier. Typically, concrete and specially rolled HDPE plastic are the materials used to create the barrier, which is placed in a 60– to 90-cm-deep ditch around the planting, and angled out at the top to direct the rhizomes to the surface. (This is only possible if the barrier is installed in a straight line.) If the containment area is small, this method can be detrimental to ornamental bamboo, as the bamboo within can become rootbound and start to display the signs of any unhealthy containerized plant. In addition, rhizomes can escape over the top, or beneath the barrier if it is not deep enough. Strong rhizomes and tools can penetrate plastic barriers, so care must be taken. In small areas, regular maintenance may be the best method for controlling the running bamboos. Barriers and edging are unnecessary for clump-forming bamboos, although these may eventually need to have portions removed if they become too large.

The ornamental plant sold in containers and marketed as "lucky bamboo" is actually an entirely unrelated plant, *Dracaena sanderiana*. It is a resilient member of the lily family that grows in the dark, tropical rainforests of Southeast Asia and Africa. Lucky bamboo has long been associated with the Eastern practice of *feng shui*. On a similar note, Japanese knotweed is also sometimes mistaken for a bamboo, but it grows wild and is considered an invasive species. *Phyllostachys* species of bamboo are also considered invasive and illegal to sell or propagate in some areas of the US.

Uses

Culinary: Although the shoots (new culms that come out of the ground) of bamboo contain a toxin taxiphyllin (a cyanogenic glycoside) that produces cyanide in the gut, proper processing renders them edible. They are used in numerous Asian dishes and broths, and are available in supermarkets in various sliced forms, in both fresh and canned versions. The golden bamboo lemur ingests many times the quantity of the taxiphyllin-containing bamboo that would kill a human.

The bamboo shoot in its fermented state forms an important ingredient in cuisines across the Himalayas. In Assam, India, for example, it is called *khorisa*. In Nepal, a delicacy popular across ethnic boundaries consists of bamboo shoots fermented with turmeric and oil, and cooked with potatoes into a dish that usually accompanies rice in Nepali).

In Indonesia, they are sliced thin and then boiled with *santan* (thick coconut milk) and spices to make a dish called *gulai rebung*. Other recipes using bamboo shoots are *sayur lodeh* (mixed vegetables in coconut milk) and *lun pia* (sometimes written *lumpia*: fried wrapped bamboo shoots with vegetables). The shoots of some species contain toxins that need to be leached or boiled out before they can be eaten safely.

Pickled bamboo, used as a condiment, may also be made from the pith of the young shoots.

The sap of young stalks tapped during the rainy season may be fermented to make *ulanzi* (a sweet wine) or simply made into a soft drink. Bamboo leaves are also used as wrappers for steamed dumplings which usually contains glutinous rice and other ingredients.

Pickled bamboo shoots are cooked with black-eyed beans as a delicacy food in Nepal. Many Nepalese restaurant around the world serve this dish as *aloo bodi tama*. Fresh bamboo shoots are sliced and pickled with mustard seeds and turmeric and kept in glass jar in direct sunlight for the best taste. It is used alongside many dried beans in cooking during winter months. Baby shoots (Nepali: tusa) of a very different variety of bamboo native to Nepal is cooked as a curry in Hilly regions.

In Sambalpur, India, the tender shoots are grated into juliennes and fermented to prepare *kardi*. The name is derived from the Sanskrit word for bamboo shoot, *karira*. This fermented bamboo shoot is used in various culinary preparations, notably *amil*, a sour vegetable soup. It is also made into pancakes using rice flour as a binding agent. The shoots that have turned a little fibrous are fermented, dried, and ground to sand-sized particles to prepare a garnish known as *hendua*. It is also cooked with tender pumpkin leaves to make sag green leaves.

The empty hollow in the stalks of larger bamboo is often used to cook food in many Asian cultures. Soups are boiled and rice is cooked in the hollows of fresh stalks of bamboo directly over a flame. Similarly, steamed tea is sometimes rammed into bamboo hollows to produce compressed forms of Pu-erh tea. Cooking food in bamboo is said to give the food a subtle but distinctive taste.

In addition, bamboo is frequently used for cooking utensils within many cultures, and is used in the manufacture of chopsticks. In modern times, some see bamboo tools as an ecofriendly alternative to other manufactured utensils.

Medicine: Bamboo is used in Chinese medicine for treating infections and healing.

Construction: Bamboo, like true wood, is a natural composite material with a high strength-to-weight ratio useful for structures.

In its natural form, bamboo as a construction material is traditionally associated with the cultures of South Asia, East Asia and the South Pacific, to some extent in Central and South America, and by extension in the aesthetic of Tiki culture. In China and India, bamboo was used to hold up simple suspension bridges, either by making cables of split bamboo or twisting whole culms of sufficiently pliable bamboo together. One such bridge in the area of Qian-Xian is referenced in writings dating back to 960 AD and may have stood since as far back as the third century BC, due largely to continuous maintenance.

Bamboo has also long been used as scaffolding; the practice has been banned in China for buildings over six storeys, but is still in continuous use for skyscrapers in Hong Kong. In the Philippines, the nipa hut is a fairly typical example of the most basic sort of housing where bamboo is used; the walls are split and woven bamboo, and bamboo slats and poles may be used as its support. In Japanese architecture, bamboo is used primarily as a supplemental and/or decorative element in buildings such as fencing, fountains, grates and gutters, largely due to the ready abundance of quality timber.

Various structural shapes may be made by training the bamboo to assume them as it grows. Squared sections of bamboo are created by compressing the growing stalk within a square form. Arches may similarly be created by forcing the bamboo's growth into the desired form, costing much less than it would to obtain the same shape with regular wood timber. More traditional forming methods, such as the application of heat and pressure, may also be used to curve or flatten the cut stalks.

Bamboo can be cut and laminated into sheets and planks. This process involves cutting stalks into thin strips, planing them flat, and boiling and drying the strips; they are then glued, pressed and finished. Long used in China and Japan, entrepreneurs started developing and selling laminated bamboo flooring in the West during the mid-1990s; products made from bamboo laminate, including flooring, cabinetry, furniture and even decorations, are currently surging in popularity, transitioning from the boutique market to mainstream providers such as Home Depot. The bamboo goods industry (which also includes small goods, fabric, etc.) is expected to be worth $25 billion by 2012. The quality of bamboo laminate varies among manufacturers and varies according to the maturity of the plant from which it was harvested (six years being considered the optimum); the sturdiest products fulfill their claims of being up to three times harder than oak hardwood while others may be softer than standard hardwood.

Bamboo intended for use in construction should be treated to resist insects and rot. The most common solution for this purpose is a mixture of borax and boric acid. Another process involves boiling cut bamboo to remove the starches that attract insects.

Bamboo has been used as reinforcement for concrete in those areas where it is plentiful, though dispute exists over its effectiveness in the various studies done on the subject. Bamboo does have the necessary strength to fulfil this function, but untreated bamboo will swell with water absorbed from the concrete, causing it to crack. Several procedures must be followed to overcome this shortcoming.

Several institutes, businesses, and universities are researching the use of bamboo as an ecological construction material. In the United States and France, it is possible to get houses made entirely of bamboo, which are earthquake- and cyclone-resistant and internationally certified. In Bali, Indonesia, an international K-12 school, the Green School, is constructed entirely of bamboo, for its beauty and advantages as a sustainable resource. There are three ISO standards for bamboo as a construction material.

In parts of India, bamboo is used for drying clothes indoors, both as a rod high up near the ceiling to hang clothes on, and as a stick wielded with acquired expert skill to hoist, spread, and to take down the clothes when dry. It is also commonly used to make ladders, which apart from their normal function, are also used for carrying bodies in funerals. In Maharashtra, the bamboo groves and forests are called Veluvana, the name *velu* for bamboo is most likely from Sanskrit, while *vana* means forest.

Furthermore, bamboo is also used to create flagpoles for saffron-coloured, Hindu religious flags, which can be seen fluttering across

India, especially in Bihar and Uttar Pradesh, as well as in Guyana and Suriname in South America.

Bamboo was used for the structural members of the India pavilion at Expo 2010 in Shanghai. The pavilion is the world's largest bamboo dome, about 34 m (112 ft) in diametre, with bamboo beams/members overlaid with a ferro-concrete slab, waterproofing, copper plate, solar PV panels, a small windmill, and live plants. A total of 30 km (19 mi) of bamboo was used.

The dome is supported on 18-m-long steel piles and a series of steel ring beams. The bamboo was treated with borax and boric acid as a fire retardant and insecticide and bent in the required shape. The bamboo sections were joined with reinforcement bars and concrete mortar to achieve the necessary lengths.

Textiles: Since the fibres of bamboo are very short (less than 3 mm (0.12 in)), they are not usually transformed into yarn by a natural process. The usual process by which textiles labelled as being made of bamboo are produced uses only rayon made from the fibres with heavy employment of chemicals.

To accomplish this, the fibres are broken down with chemicals and extruded through mechanical spinnerets; the chemicals include lye, carbon disulfide and strong acids. Retailers have sold both end products as "bamboo fabric" to cash in on bamboo's current ecofriendly cachet; however, the Canadian Competition Bureau and the US Federal Trade Commission, as of mid-2009, are cracking down on the practice of labelling bamboo rayon as natural bamboo fabric. Under the guidelines of both agencies, these products must be labelled as rayon with the optional qualifier "from bamboo".

Paper: Bamboo fibre has been used to make paper in China since early times. A high-quality, handmade paper is still produced in small quantities. Coarse bamboo paper is still used to make spirit money in many Chinese communities.

Bamboo pulps are mainly produced in China, Myanmar, Thailand and India, and are used in printing and writing papers. The most common bamboo species used for paper are *Dendrocalamus asper* and *Bamboo bluemanea.* It is also possible to make dissolving pulp from bamboo. The average fibre length is similar to hardwoods, but the properties of bamboo pulp are closer to softwood pulps due to it having a very broad fibre length distribution. With the help of molecular tools, it is now possible to distinguish the superior fibre-yielding species/varieties even at juvenile stages of their growth, which can help in unadulterated merchandise production.

Musical Instruments

Other uses: Bamboo has traditionally been used to make a wide range of everyday utensils, particularly in Japan, where archaeological excavations have uncovered bamboo baskets dating to the Late Jomon period (2000-1000 BC).

Bamboo has a long history of use in Asian furniture. Chinese bamboo furniture is a distinct style based on a millennia-long tradition.

Several manufacturers offer bamboo bicycles, surfboards, snowboards, and skateboards.

Due to its flexibility, bamboo is also used to make fishing rods. The split cane rod is especially prized for fly fishing. Bamboo has been traditionally used in Malaysia as a firecracker called a *meriam buluh* or bamboo cannon. Four-foot-long sections of bamboo are cut, and a mixture of water and calcium carbide are introduced. The resulting acetylene gas is ignited with a stick, producing a loud bang. Bamboo can be used in water desalination. A bamboo filter is used to remove the salt from seawater.

Food is cooked in East Timor in bamboo in fire. This is called *Tukir*.

Many minority groups in remote areas that have water access in Asia use bamboo that is 3–5 years old to make rafts. They use 8 to 12 poles, 20 to 24 feet long, laid together side by side to a width of about 3 to 4 feet. Once the poles are lined up together, they cut a hole crosswise through the poles at each end and use a small bamboo pole pushed through that hole like a screw to hold all the long bamboo poles together. Floating houses use whole bamboo stalks tied together in a big bunch to support the house floating in the water. Bamboo is also used to make eating utensils such as chop sticks, trays, and tea scoops.

In Asian Culture

Bamboo's long life makes it a Chinese symbol of upright, while in India it is a symbol of friendship. The rarity of its blossoming has led to the flowers' being regarded as a sign of impending famine. This may be due to rats feeding upon the profusion of flowers, then multiplying and destroying a large part of the local food supply. The most recent flowering began in May 2006. Bamboo is said to bloom in this manner only about every 50 years.

Bamboo, Noble and Useful

Bamboo, one of the "Four Gentlemen" (bamboo, orchid, plum blossom and chrysanthemum), plays such an important role in traditional Chinese culture that it is even regarded as a behaviour model of the gentleman.

As bamboo has some features such as uprightness, tenacity and hollow heart, people endow bamboo with integrity, elegance and plainness, though it is not physically strong. Ancient Chinese poets wrote countless poems to praise bamboo, but actually they were truly talking about people like bamboo and express their understanding of what a real gentleman should be like.

According to laws, an ancient poet, Bai Juyi (772–846), thought that to be a gentleman, a man does not need to be physically strong, but he must be mentally strong, upright, and perseverant. Just as a bamboo is hollow-hearted, he should open his heart to accept anything of benefit and never have arrogance or prejudice. Bamboo is not only a symbol of gentleman, but also an important role in Buddhism.

In the first century, Buddhism was introduced into China. As canons of Buddhism do not allow its believers to do anything cruel to animals, flesh and egg were not allowed in the diet. However, people need something nutritious to live; thus, the tender bamboo shoot (*sÔn*M{ in Chinese) became a good choice. The bamboo shoot is nutritious, and eating it does not violate the canon. With thousands of years' development, how to eat bamboo shoot has become a part of cuisine system, especially for monks.

A Buddhist monk, Zan Ning, wrote a manual of the bamboo shoot called "*SÔn PÔ*M{\‹". He offered descriptions and recipes for many kinds of bamboo shoots. Bamboo shoot has always been a traditional dish on the Chinese dinner table, especially in southern China. In ancient times, as long as people had money to buy a big house with yard, they would always plant bamboos in their garden. Bamboo is a necessary element of Chinese culture, or even in the whole Asian civilization. People plant bamboos, eat bamboo shoots, paint bamboos, write poems for bamboos, and speak highly of gentlemen who are like bamboos. Bamboo is not only a plant, but also a part of people's lives.

In Japan, a bamboo forest sometimes surrounds a Shinto shrine as part of a sacred barrier against evil. Many Buddhist temples also have bamboo groves.

In northern Indian state of Assam, the fermented bamboo paste known as *khorisa* is known locally as a folk remedy for the treatment of impotence, infertility, and menstrual pains.

Bamboo plays an important part of the culture of Vietnam. Bamboo symbolizes the spirit of Vovinam (a Vietnamese martial arts): *cýõng nhu phÑi triÃn* (coordination between hard and soft (martial arts)). Bamboo also symbolizes the Vietnamese hometown and Vietnamese soul: the gentlemanlike, straightforwardness, hard working, optimism, unity, and adaptability. A Vietnamese proverb says, "When the bamboo is old, the bamboo sprouts appear", the meaning being Vietnam will never be annihilated; if the previous generation dies, the children take their place. Therefore, the Vietnam nation and Vietnamese value will be maintained and developed eternally. Traditional Vietnamese villages are surrounded by thick bamboo hedges (*liy tre*).

The Song Dynasty (960–1279 AD) Chinese scientist and polymath Shen Kuo (1031–1095) used the evidence of underground petrified bamboo found in the dry northern climate of Yan'an, Shanbei region, Shaanxi province to support his geological theory of gradual climate change.

In Mythology

Several Asian cultures, including that of the Andaman Islands, believe humanity emerged from a bamboo stem.

In Philippine mythology, one of the more famous creation accounts tells of the first man, *Malakás* ("Strong"), and the first woman, *Maganda* ("Beautiful"), each emerged from one half of a split bamboo stem on an island formed after the battle between Sky and Ocean. In Malaysia, a similar story includes a man who dreams of a beautiful woman while sleeping under a bamboo plant; he wakes up and breaks the bamboo stem, discovering the woman inside. The Japanese folktale "Tale of the Bamboo Cutter" (*Taketori Monogatari*) tells of a princess from the Moon emerging from a shining bamboo section. Hawaiian bamboo *('ohe)* is a *kinolau* or body form of the Polynesian creator god Kâne.

A bamboo cane is also the weapon of Vietnamese legendary hero, Saint Giong, who had grown up immediately and magically since the age of three because of his wish to liberate his land from Ân invaders. An ancient Vietnamese legend (The Hundred-knot Bamboo Tree) tells of a poor, young farmer who fell in love with his landlord's beautiful daughter. The farmer asked the landlord for his daughter's hand in marriage, but the proud landlord would not allow her to be bound in marriage to a poor farmer.

The landlord decided to foil the marriage with an impossible deal; the farmer must bring him a "bamboo tree of 100 nodes". But Buddha (*Båt*) appeared to the farmer and told him that such a tree could be made from 100 nodes from several different trees. *Båt* gave to him four magic words to attach the many nodes of bamboo: *Kh¯c nh-p, kh¯c xu¥t*, which means "joined together immediately, fell apart immediately". The triumphant farmer returned to the landlord and demanded his daughter. Curious to see such a long bamboo, the landlord was magically joined to the bamboo when he touched it, as the young farmer said the first two magic words. The story ends with the happy marriage of the farmer and the landlord's daughter after the landlord agreed to the marriage and asked to be separated from the bamboo.

In a Chinese legend, the Emperor Yao gave two of his daughters to the future Emperor Shun as a test for his potential to rule. Shun passed the test of being able to run his household with the two emperor's daughters as wives, and thus Yao made Shun his successor, bypassing his unworthy son. Later, Shun drowned in the Xiang River. The tears his two bereaved wives let fall upon the bamboos growing there explains the origin of spotted bamboo. The two women later became goddesses.

As a Writing Surface

Bamboo was in widespread use in early China as a medium for written documents. The earliest surviving examples of such documents, written in ink on string-bound bundles of bamboo strips (or "slips"), date from the fifth century BC during the Warring States period. However, references in earlier texts surviving on other media make it clear that some precursor of these Warring States period bamboo slips was in use as early as the late Shang period (from about 1250 BC).

Bamboo or wooden strips were the standard writing material during the Han dynasty, and excavated examples have been found in abundance. Subsequently, paper began to displace bamboo and wooden strips from mainstream uses, and by the fourth century AD, bamboo slips had been largely abandoned as a medium for writing in China. Several paper industries are surviving on bamboo forests. Ballarpur (Chandrapur, Maharstra) paper mills use bamboo for paper production.

As a Weapon

Bamboo is used in several East Asian and South Asian martial arts.

- In the ancient Tamil martial art of *Silambam*, fighters would hit each other rapidly with bamboo sticks.
- In the Japanese martial art *Kendo*, bamboo is used to make the *Shinai* sword.
- Bamboo is used for crafting the world's longest bows, the *Yumi*, and the arrows used in the Japanese martial art *Kyudo*.
- In traditional archery, the longbow and recurve bow limbs are sometimes crafted with flat ground bamboo, and also used to make superior weapons for bowhunting and target archery.

- A bamboo stick can be made into a simple spear by sharpening one of the ends.
- History's first gunpowder-based weapons, such as the fire lance, were made of bamboo.

Environmental Plantations

Forest plantations are not anything new. In southwestern France, for example, there are records of man-made forests as early as year 1500, as well as evidence of trading in forest products, both wood and non-wood from maritime pine (Pinus pinaster) plantations. These early plantations resulted mainly from the initiative of private landlords as well as smallholders and communities, primarily for commercial purposes (i.e., the trade of wood and resin products), but also for direct household needs. In this same area there are records of zones planted with pines and other vegetation to control the advance of sand dunes before 1600. Therefore, there is evidence of commercial, smallholder and community man-made forests, for both production and protection purposes, from as early as 400 to 500 years ago.

While for all economic agents involved forest plantations constitute an investment (i.e., the application of resources in an effort to attain a greater benefit in the future), different agents will invest their resources with different purposes in mind.

For example, a private enterprise will emphasize commercial benefits, a rural household will probably be more interested in satisfying more direct needs for fuelwood and fodder, and national or local governments may focus on controlling erosion, restoring watersheds, and other types of public interest investments. One way or another, forest plantations constitute a form of land use, in which land may be competing with other possible uses, and the application of other resources, such as capital, labour and know how, which also compete against other investment opportunities.

Regardless of the type of benefit expected (whether commercial, social or other), the development of forest plantations involves the creation of fixed physical capital. Like any investment of this type, planted forests require sacrificing current consumption, so that a capability that combines factors of production to produce goods and services in the future can be established. Apart from the standard features of capital investments, forest plantations have some specific characteristics. As a form of land-use, there is no opportunity for change over a long period of time, and returns on investment are delayed for several years.

Before focusing upon the economics of forest plantations as an investment, it may be a good idea to reflect upon some technical aspects. While it may not be immediately apparent, plantation forestry involves a considerable number of technical factors that will affect the economic and social performance of a forest plantation programme. Not only do most of these factors represent economic costs, but also their proper combination and handling require specific capabilities and know-how, which are also part of the investment effort.

Basic to the development of any forest plantation programme is the availability of a sufficient area of suitable land for lease, purchase or change in land-use. Whether this land is owned by private individuals or entities, by local communities, or by the government, public authorities promoting the programme must provide for the adequate structure of rights and obligations that will enable large areas to be committed to a particular form of land use over long periods of time. A forest plantation programme will also require critical amounts of manpower at the professional, specialized and unskilled levels. Additionally, important volumes of funds must be consistently available over several years.

Above and beyond the basic factors of land, labour and funding, many other factors, most of them related to plantations as a production process must be taken into account, particularly since they will affect the economic success or failure of plantation operations. Early in the development of a plantation programme, serious consideration must be given to the choice of species and provenances, and their proper matching with sites. This site-species matching must not focus only on needs for commercial wood, but must also take account of the needs of local people and site protection and conservation requirements. In this species-site matching exercise, opportunities for combining tree planting with other forms of land-use (e.g., agroforestry) must also be examined. Part of this site-species matching work requires some technical analyses such as an understanding of growth rates, conditions and patterns, and possible rotations of the main candidate species.

Plantation forestry involves the establishment and management of nurseries, which constitute a production system in its own right, with its own set of inputs, outputs and management requirements. Nurseries must be able to provide the right amount of plants, of the right species, of the right quality, at the right time, and at the right place. Among the essential factors of nursery success is the procurement of quality seeds in the right amounts, which must be properly stored and treated.

After seedlings are grown in nurseries, planting proper takes place, which is normally the most costly and critical operation among plantation forestry activities. To ensure good survival and early growth, major costs are incurred in site preparation, drainage, initial fertilization, and weeding. Other critical elements at planting time that will affect future plantation performance are quality seedlings, the timing of planting and proper spacing.

Once the early survival of a plantation stand is assured, treatment operations involving new costs continue in the form of thinning, and perhaps pruning and some additional fertilization. Also, throughout the life of a plantation, management must provide for protection against hazards (e.g., fire, diseases and pests), and for the maintenance of machinery, equipment and infrastructure permitting the continuing management of the plantation operation.

Finally, at the end of the rotation, harvesting systems, including road and skid trail layout, as well as felling techniques must be implemented to minimize site damage in preparation for the next rotation of tree crops. Decades of experience have shown that forest plantation programmes seldom end with the first rotation, and in looking forward to successive rotations, such programmes must remain mindful of maintaining site quality through cultural treatments, and invest in research efforts focusing on trials for testing species/ provenances, responses to various silvicultural treatments, tree improvements and protection against diseases and pests.

As can be seen from this very brief description of the technical aspects of plantation forestry, this is not a simple business. On the contrary, it requires important costs, significant amounts of specific know-how, and involves several factors of risk of failure. Additionally, since the benefits derived from such efforts are not obtained until after several years, most economic agents will require certain conditions before embarking into this type of investment. It is important to stress, however, that as previously mentioned, there are tens of millions of hectares of plantation forests in the world today, which means that various economic agents over several decades and even centuries have invested in forest plantations expecting future benefits. Some of these investments have failed, but it is fair to say that perhaps most have been successful, as evidenced by the continued expansion of forest plantation activities in the world today.

While there is a recent emphasis on agroforestry and other forms of social forestry schemes, as well as plantations for watershed protection and other resource conservation purposes, most forest plantations throughout history have been driven by production or commercial motives. Various countries have designed systems of incentives (mostly fiscal and financial) to enhance forest plantations, but the expectation of commercial gain has remained the major motivating force behind most man-made forest efforts. This implies that a fundamental consideration in launching or strengthening a forest plantation programme is the access to a market where the proceeds of a plantation can be sold at a competitive price.

What would be the conditions (at least in a theoretical sense) under which an economic agent would be willing to undertake an investment in plantation forestry? First and foremost, economists assume that the agent who plants trees on a hectare of land owns or leases the right to exclusive collection of the proceeds of that which he has planted. In return for this exclusive right the agent expects to pay a land rent. If he owns the land outright, he will pay this rent to himself at an opportunity cost rate determined in a competitive market. If he leases the land he will pay the rent to a private owner or to the government at a competitive price (Samuelson 1976).

Apart from the exclusive rights to the land and its proceeds obtained at a competitive price, the economic agent is assumed to know about certain features of the markets in which he operates. These assumptions include:

- Knowledge of the future timber prices at which all outputs can be freely sold, and future prices of all inputs;
- Knowledge of future interest rates at which the agent can both borrow and lend in indefinite amounts;
- Knowledge of technical timber yields that emerge at future dates once certain expenditure inputs are made (planting, fertilization, thinning, pruning, and other cultural treatments);
- Finally it is assumed that each kind of land suitable for forest plantations can be bought and sold and rented at arm's length transactions between numerous competitors; or, if the government owns public lands, it rents them out at auction to the highest of numerous alternative bidders (Samuelson 1976).

Under these assumptions, it can be proven that the agent can determine an optimum rotation period, which results from maximizing the present discounted value of all net receipts (excluding land rents) calculated over an infinite chain of cycles, or what turns out to be the same, maximizing the present discounted value of net receipts over the first cycle, but with the market land rental included in those receipts. Additionally, this is theoretically a sustainable solution (Samuelson 1976). This is essentially the optimal solution developed by the German forester Martin Faustmann in 1849, which has been supported over time by several authors as theoretically correct.

An important point in this theoretical approach is that the present discounted value calculation implies the existence of a positive interest rate, which clearly militates against long-lived investment projects. Compound interest is indeed the legitimate return that economic agents must earn in forest plantations if they are not to employ their land, labour and available funds in other more lucrative uses.

One major lesson from this theoretical approach is that forest plantations are economically no different from other investment opportunities, and that wood is economically no different from other commodities transacted in competitive markets. As wood becomes scarce, it will become more expensive. As long as there remain important needs for wood that people will want to satisfy, the price of wood will rise to the level necessary to keep a viable supply of it forthcoming (Samuelson 1976). This is where forest plantations come in as part of the solution to a problem of wood scarcity.

Yes, Professor Samuelson's framework focuses strictly on the production aspect of plantation forestry, but most of us know (including Samuelson) that land use is subject to all kinds of externalities, and as a result, many land-use problems are often handled through regulatory processes. Nevertheless, on the basis of decades of experience, there is reason to believe that private economic agents motivated by the expectation of financial returns can effectively handle a major part of forest plantation efforts.

What about forest plantations at the smallholder or village level? While in these cases the emphasis is not normally on the selling of commodities to maximize net returns based on commercial transactions, small farmers are also economic agents attempting to maximize their well-being. Most often, smallholder and village plantations are undertaken to provide for direct needs such as fuelwood, fodder, fencing, food (as in the case of fruit trees) and as shade and protection of agricultural crops (e.g., shelterbelts).

However there is also evidence that under certain conditions, smallholders do plant trees in anticipation of future markets, implying that even under social forestry programmes, the access to markets must not be neglected. A basic condition for these plantations to occur appears to be the security of rights, as mentioned above under the assumptions of the theoretical approach. Another, more practical condition, seems to be the provision of seedlings (often at subsidized prices) through governmental extension services or NGOs.

What about forest plantations for protection and conservation purposes? There is ample evidence that in tropical countries the loss of forest cover has led to extensive damage particularly derived from soil erosion. This is largely due to climatic factors (such as prolonged heavy rains, major changes in relative humidity and high temperatures) and the inherent fragility of tropical soils, but also to land use practices such as periodic burning of grasslands.

While these problems have long been recognized, little protective planting has been performed, and much of what has been done has failed, although the reasons for failure are rarely technical (Evans 1992). The complexity and interdisciplinary nature of conservation efforts associated with forest plantations require coordination of several government agencies at national, regional and local levels. Additionally, effective watershed management involves not only forestry, but also agriculture, hydrology, engineering, and social capabilities related to the participation of the local people. The coordination of these many disciplines requires a comprehensive organization lacking in many developing countries (Evans 1992).

Regardless of whether the results will be commercial forests, agroforestry activities, or watershed management/conservation efforts, a plantation forestry programme will constitute a major investment of land, human, material and financial resources. Additionally, while the bulk of the programme may end up in commercial plantations conducted and managed by the private sector, the State will have to play an active role (not only in setting policies and providing a leadership and coordinating role, but also as a direct agent), and so will local governments and NGOs.

The role of the State as the key actor in the design of a national forest plantation programme and in creating the enabling policy conditions for its implementation is fairly obvious. However, State agencies will also have to play a more direct role in specific activities for which the private sector will not face the motivation of the markets. Protective plantations, for example, while economically important through their effect on environmental and resource conservation, do not have a direct financial return, and therefore, either the State or other public agents will have to assure the implementation and management of such investments. Additionally, under social forestry and agroforestry schemes, either directly, or through local governments or NGOs, the State will have to ensure an adequate supply of seedlings for a variety of situations, as well as technical services to communities and smallholders to ensure the success of activities in the field. Another role for government agencies at various levels would be the coordination of efforts to protect forest plantations against hazards, particularly fire. Finally, a forest plantation programme with a long-term vision must be supported by a research effort that will ensure continuous improvement in the quality of planting stock, in the proper matching of species/provenances and sites, in soil management and maintenance and in optimizing stand treatments.

In sum, while the private sector is expected to play a major role, a necessary condition is that private agents need to be convinced of expecting an economic advantage from committing large amounts of resources over long periods of time to the development of forest plantations. This will depend on a number of factors that will undergo further review in the next section. Smallholders and communities will also need to see the advantage in taking a long-term perspective. Finally, the State will also have to play an active role and commit resources to those aspects of the programme, which while economically important, will not have direct financial returns and may require a continuing commitment of public funds.

A Simple Economic Model

A simple representation of a landscape with two forms of land-use, namely agriculture and forestry. This simplified model will help us understand under what conditions of markets and policies, and on what types of land, economic agents will be interested in investing in forest plantations. The vertical axis measures the value of land as a productive asset, while the horizontal axis represents distance from, or access to markets. Before going any further, it must be noted that this model is based on the premises of an economy of competitive markets, as discussed in the previous section, where the structure of rights and obligations of economic agents is clearly defined.

Agricultural land is represented by the value function extending from the vertical axis to point D. The value of farmland is highest at the intercept of the vertical axis, and it declines with increasing distance or decreasing access to the market, which is represented by the origin. Forest land is shown by the value function extending from the vertical axis to point C. Additionally, since land ownership involves some costs of excluding others and claiming its benefits privately, the upward sloping function c_r shows the marginal cost of establishing and maintaining secure rights to the land. While the cost of obtaining legal title may remain constant with distance or access, the cost of excluding trespassers may increase substantially up to the point that no security effort, no matter how costly, will exclude illegal users in remote areas.

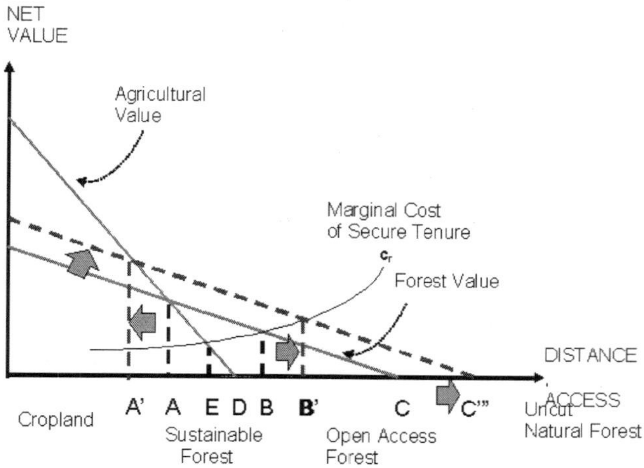

Figure: *The landscape of forest development*

The agricultural land-use value function and the forest land-use value function intersect above point A. From the origin to this point, since agricultural values are higher than forest values, the natural forest has been cleared over time to make room for agricultural activities. To the right of point A, however, forest values are higher than agricultural values, and land from point A to point C will be used to obtain benefits from forest products. The area to the right of point C remains as unharvested forest because the costs of its utilization exceed the value of the products removed.

So far, nothing has been said about the sustainability of either of the two forms of land-use considered. A major factor in sustainable land use will be the cost of securing property rights represented by the curve c_r. The functions showing agricultural land value and the cost of secure property intersect above point E. This implies that, in the absence of a competing forest value, farmers would be interested in investing and managing sustainable agriculture between the origin and point E.

Under these same conditions (i.e., no competing forest value), the area between points E and D (where agricultural land value is zero) will be treated as an open access resource to be exploited for short-term advantage. Similarly, forest users will be interested in investing and practicing sustainable forest management in the area extending between points A and B, while the area between points B and C (where forest land value is zero), will remain as an open access forest, which will be left to regenerate without investments and harvested periodically for short term gain. However, as previously discussed, forest use begins to compete favourably with agriculture to the right of point A. Under the conditions displayed, therefore, it can be expected that there will be sustainable agriculture from the origin to point A, sustainable forestry from point A to point B, no unsustainable agriculture in any significant amount, open access (unsustainable) forest use from point B to point C, and uncut natural forest to the right of point C.

An important conclusion that can be drawn from examining this model, is that under conditions of competitive markets and clearly established usage rights, sustainable forest management can be practiced by private economic agents in areas where it provides long-term economic advantages if the cost of establishing and maintaining such rights is included.

However, no sustainable management can be expected in areas where the increasing cost of maintaining usage rights precludes investments in maintaining the resource. From the standpoint of forest plantations, which are perhaps the most important form of sustainable forestry nowadays, these can be expected to occur in some areas where the natural forest has been logged or otherwise degraded, but definitely not in all.

In the beginning of this section, it was mentioned that this simplified model could help understand conditions of markets and policies that would affect the interest of economic agents in undertaking forest plantations. This can be illustrated by shifting either the value functions or the tenure cost function. The dashed line extending from the vertical axis to point C' represents a shift to the right in the forest value function (i.e., an increase in forest value). Such a shift can come about either through market forces without any policy intervention (like an increase in demand for wood products over time), through the opening of new markets (such as a new policy of international trade facilitation), or even through policies that favour forest land value (e.g., special fiscal treatments).

Such a shift would have two important effects. First, there would be an expansion of the area in which economically profitable sustainable forest management can be practiced (i.e., the area between points A and B expands to the area between points A' and B'). This expansion would derive from a reduction in the area of sustainable cropland (i.e., the area between points A and A'), on which forestry would now be more profitable than agriculture, and from a reduction of the area of open access forests (i.e., the area between points B and B') in which it has now become profitable to invest in sustainable forest management. Second, an expansion of the area of open access forest (i.e., the area extending from point C to point C') can also be expected as it now becomes profitable to cut into new forest areas previously left uncut.

While not shown in order to avoid overcrowding Figure 1, or to avoid overcrowding this paper with additional figures, this type of model can be used to examine the effects of several other changes, of which we can give a few examples:

1. The forest value function could shift to the left, either as a result of a decline in forest product prices or the production of goods from the forest becoming more expensive as a result of restrictive policies. The effect would be a reduction in the area of sustainable forest management, and an increase in the area of sustainable farming.
2. The agricultural value function could shift to the right, as a result, for example of a policy of subsidized prices for certain crops. The effect would be an increase of the area under sustainable agriculture and a reduction in the area under sustainable forestry.
3. The function representing the cost of secure tenure could shift to the right. This could come about under land reform policies that would include the privatization of certain public lands. This could result in increases of land under both sustainable farming and sustainable forestry.
4. The agricultural value function could shift to the left, perhaps as a result of a drop of international prices of certain cash crops. The effect would be a reduction of the area sustainably farmed and an increase in the area under sustainable forestry.

There are many other possibilities: while the above discussion considered the shift of one of the functions in the diagram at a time, it is perhaps more common in the real world that several of these shifts are taking place simultaneously, thus affecting in many ways the economic opportunities for the establishment of forest plantations. Policy changes such as fees, taxes, forest charges, subsidies, regulatory restrictions, incentives favouring particular forms of land use, etc., all trace their effects through these shifts.

At least three important lessons can be learned from this simple model. First, as mentioned earlier in this section, economic agents, as they attempt to maximize their well-being, will show interest in investing in forest plantations under conditions that will make them expect an economic advantage with reasonable certainty. Not all land previously covered with forests will satisfy these conditions. Some will be dedicated to sustainable forms of land use with higher economic benefits than forestry (e.g. sustainable agriculture or urban development); some other areas will support forest plantations, and perhaps other forms of sustainable forestry; and finally some will remain as open access land where no sustainable use can be expected. Second, the area in which investment in forest plantations can be expected is by no means permanent. As discussed, this area will vary as a result of market forces affecting land use margins, even in the absence of policy changes. Third, the policy environment can make a major difference in the extent of the area in which sustainable forestry will be practiced. Public policies can be of several kinds, from macroeconomic measures and broad-based institutional reforms affecting the whole economy, to very specific instruments and prescriptions focusing on particular sectors. The economic agent, especially when facing opportunities for long term investments will react favourably to a stable policy environment, and therefore policy options must be analyzed in an integrated fashion, with explicit consideration of how they interact with each other.

It is clear that the simple model has focused on private agents expecting economic gains from commercial transactions derived from the utilization of forest resources. However, some may ask, what about forest plantations carried out by communities and smallholders, and what about protection forests? Well, that is why we have consistently referred to this diagram as a simple model, i.e., in order to focus on certain aspects of plantation forestry we have left some out. The reason we focused on private economic agents is simply because most of the areas reforested in the world has been driven by economic agents expecting some kind of direct economic return from their investment. Smallholders, however, are often integrated into some kind of market for forest products. In an informal way, they carry out a similar kind of economic reflection to sort out trade-offs, and by the way, this kind of informal economic thinking is also performed if the focus is on direct household needs, i.e., what will the household gain (in terms of shade, forage, fuelwood, etc.), from the time and effort applied to planting trees, even if the seedlings are subsidized?

Regarding forest plantations for protection purposes, these are not represented in the model, simply because private economic agents will rarely embark in substantial investments that do not have a direct return, regardless of how beneficial such investments may be for the society at large. While private agents can be expected to be economically more efficient in investments leading to marketable outputs, this efficiency is unlikely for non-marketed services like environmental benefits. Forest plantations for environmental protection purposes are investments of a more social nature, not too different from road infrastructure, or public education, and they are often conducted by some type of public authority and with public funding. However, these are investments nonetheless, and they are conducted in the interest of generating environmental protection advantages for society. While markets will not directly provide information to assess the values of such advantages, they can be estimated indirectly, and therefore it is possible to conduct calculations that represent the economic efficiency of investments in protection plantations.

Protection plantations are likely to occur in the area extending from point B to point C, i.e., the open access forest, especially if there is a risk of environmental damages, such as erosion or flooding that certain economic agents must endure through no choice of their own (i.e., externalities). If investments of this kind are to be economically efficient, public authorities, just like private economic agents will need to apply economic calculations to ensure that the costs do not exceed the benefits. It must be remembered that inefficiency in public investments results in a loss to society.

While most of the discussion in this section revolved around a simple diagrammatic model, a major intent was to show the effect of economic forces and policies on opportunities for investing in forest plantations. The importance of a stable policy environment, especially when facing long-term investments, cannot be overemphasized. Investors need to know what to expect, including overall sociopolitical and institutional stability, macroeconomic policies, overall environmental policies, access to markets, tenure rights, and regulations specific to the forestry and forest products sector.

As mentioned in Section II, interest rates militate against long-lived investment projects, but an unstable policy environment is probably an even more negative factor. In the presence of macroeconomic instability and policy uncertainty, economic agents will underinvest in plantations because they cannot be sure of obtaining the full return on their investments before the economic or policy environment turns to their disadvantage (Hyde *et al.* 1996).

In terms of Figure 1, an unstable policy environment could mean that economic agents do not have confident expectations about whether government rules defining c_r (the function showing costs of secure land use rights) will either relax or become more confining. Such environment could also mean that the agent does not know whether government log marketing requirements will relax or tighten. Under such uncertainty, firms will behave as if c_r and the forest land value function were to the left of where they actually are, in order to protect themselves against potentially unfavourable policy changes. This would result in less land under sustainable forest management than if the policy environment was stable.

Finally, as the needs and aspirations of a society evolve over time, institutions and policies must adjust. This brings up the importance of open and participatory deliberative processes, involving the different sectors of a society, so that policy and institutional changes can be implemented with enough transparency for expectations about change to remain consistent.

A Case: Two Hundred Years of Sustainable Forest Plantations

In the introductory section, it was mentioned that the "Landes de Gascogne" in Southwestern France would provide examples given its long history of forest plantation management. While nowadays, France cannot be considered a developing country, one of the reasons for selecting this region is that it has shown over centuries that forest plantations are and can remain an economically viable and sustainable land-use option. Additionally, from the standpoint of economic development, during most of the nineteenth century, as France was slowly but surely becoming an industrialized nation, the Landes was a comparatively underdeveloped region, which owes most of its current prosperity to forest plantations.

By far, the most important species in those million hectares of forest plantations is the maritime pine (Pinus pinaster), and therefore the "Forêt landaise" is clearly a monoculture, although it has proven to be sustainable, at least over more than two hundred years. This is largely because the soils in the Landes are poor, and this species is well adapted to such conditions. Additionally, from a commercial standpoint, Pinus pinaster has proven to be versatile enough in that it has provided transformation activities over time with raw materials that have adapted well to changing markets.

The million hectares of the Landes forest, the total area of which has remained virtually unchanged over more than 100 years, is 93% private, with more than 40,000 owners of all sizes, and is largely oriented towards the production of raw materials for the forest products industry. The remaining 7% is composed mostly of coastal protection forests owned by the State, as well as some community-owned forests (Lescourges 1994).

Historically, in the Landes, land-use was regulated by a well-defined structure of customary rights and obligations, both individual and community-based, which has existed at least since medieval times, with very limited intervention on the part of State authorities. Before the spectacular development of forest plantations in the nineteenth century, earlier planted forests coexisted with sheep raising, which constituted then the most important form of land use.

Two major operations conducted in the early nineteenth century favoured the rapid expansion of plantation forests in the Landes: (i) the control of coastal sand dunes through plantations of trees and other forms of vegetation, as well as mechanical techniques; and (ii) the control of superficial water tables through drainage in the interior of the region. This second operation was facilitated by a law enacted in 1857 on the drainage and afforestation of the Landes de Gascogne, even though the evolution of land-use in favour of pine plantations was already well underway as a result of the economic decline of sheepherding and the increasing access to markets for forest products.

As forest plantations (driven by increasing access to markets and other factors), started taking land away from sheepherding, complex land use conflicts developed involving communities, private owners and the State, some of which were not finally resolved until the mid-twentieth century. However, forest plantations finally prevailed, and by the end of the nineteenth century, the Landes forest was already well established with its million hectares largely dominated by Pinus pinaster.

An important point must be made regarding the control of coastal dunes mentioned above. These efforts began in earnest in the early nineteenth century and for the most part have been conducted by the State as protection forestry projects in the interest of threatened coastal communities, infrastructure, wetlands, ponds and lakes, and pasturelands. These coastal forests are still State property today, and protection continues to be their main function, along with recreation.

Fire has been an ever-present problem in this region. Originally, shepherds used fire to control unpalatable vegetation invading pasturelands. During the nineteenth century, however, the above-mentioned land use conflicts often resulted in deliberate arson against pine plantations. More recently, especially in the mid-to late 1940s, major fires have burnt large areas due in part to insufficient fire control programmes. Throughout the twentieth century, and especially in the second half, major efforts of forest fire control have been undertaken with the participation of the State, associations of private owners and local governments. These efforts, currently quite effective, have included fire-related infrastructure (e.g., firebreaks, access roads, and water supply), equipment, the establishment of a permanent forest firefighting force, the development of fire insurance instruments, and financial aid for reforesting damaged areas.

From a forest product standpoint, early major wood uses were lumber and charcoal making. However, resin, a non-timber forest product, was the dominant part of the early economy of forest plantations in this area, and continued to play an important economic role well into the twentieth century. Nowadays, the Landes forest supports the largest and most diversified forest products industry of any region in France, including lumber, plywood, composite panels, as well as pulp and paper, and almost all of the harvested volume is processed locally.

Given this broad-based market, the objective of the large majority of forest owners in the Landes is to attain the highest possible production of high quality wood, in the least amount of time, in harmony with maintaining the local ecosystem. An important dimension of this raw material market is that wood quality has a major price effect. For this reason, most owners are willing to incur major investments in silvicultural treatments, which consistently include site preparation, genetically improved seedlings, drainage, fertilization, weed control, pruning, and thinning, all of which are costly operations. Such investments, supported by research programmes, both public and private, have resulted in the doubling of stand productivity over the past 25 years (Arbez cited in Gelpe 1994b).

The role of the State in research and support services has been most important in the area of silviculture. Not only have State research agencies played a major role in tree improvement (particularly the establishment management of seed orchards and the certification of seed sources) and sophisticated silvicultural modelling; State agencies also play a major role in research on the effects of production forest management on the environment and the conservation of biodiversity, and in further development of sand dune control techniques.

In this day and age, the Landes de Gascogne is a prosperous and dynamic forest region, competing effectively in an open international market, which has proven over decades and even centuries of private plantation forestry, that this is a sustainable and viable land use option, provided that certain conditions are present. Among these, the following are worth mentioning:

- A plantable tree species that is well adapted to local environmental conditions, with reasonable growth and yield potential and with flexible technological characteristics that allow for adaptation to evolving markets.

- Access to open markets under fair competitive conditions.
- A stable legal, institutional and policy framework that allows private economic agents to develop predictable and consistent expectations about future opportunities.
- Government institutions and agencies, that while allowing private agents to perform freely, provide support in areas where the private sector may be weak (e.g. research and fire protection) and concentrate on land management under conditions in which private agents may be inefficient.

Finally, a brief word about reforestation incentives: France has developed, especially during the twentieth century, a number of incentive policies to enhance reforestation. As a result of such policies, France's forest cover has grown from about 11 million ha at the end of World War II, to approximately 15 million ha today. The most important among these policies have been the exoneration of property taxes enacted in 1934 and the National Forestry Fund, established in 1946 to provide financial incentives for reforestation. While these policies may have facilitated the continuity of forest plantations in this region, the Landes forest had been completely established by the turn of the century, and therefore, it did not originally benefit from either policy. This suggests that more basic economic and institutional conditions may be more important than sector-specific incentives for the development of forest plantations.

Other Cases

While their history is not as long as that of the Landes de Gascogne, there exist several other examples in different parts of the world where forest plantations have prevailed as major forms of land-use. Apart from Nordic and central European countries (e.g., Sweden, Finland and Germany) with their long forestry history, more recent, but still major lessons of experience can be found in the United States (particularly the Southeast), Brazil, Chile, China, New Zealand and others. This section will focus on experiences from Chile, New Zealand and China. All these experiences are similar in the sense that vast areas have been reforested in recent decades in all these countries, but they are also different from the standpoint of their institutional and policy contexts.

Chile: A Forest Economy based on Pinus Radiata

The large majority of Chilean forest plantations (currently a total of 1.8 million ha) are made up of Pinus radiata, which, much like Pinus pinaster in the French southwest constitutes large areas of monoculture. Although this Californian species was introduced to Chile in 1885, it was not until 1935 that forest plantations began in earnest. While an exotic, Pinus radiata has found in the coastal areas of central Chile ideal conditions of soil and climate, where it has become a major factor of the land-based economy over the past half-century. Most early plantations were established on degraded soils that could no longer support sustainable agriculture. Throughout their history, these plantations have been largely the product of private investments, in which small owners have always played a major role.

Through forest legislation introduced in the early 1930s, the State began encouraging the establishment of forest plantations, especially by means of tax exonerations, and the distribution of subsidized seedlings. However, the major force behind the early development of planted forests was the economic advantage anticipated by small owners and corporations from a fast-growing and versatile species. Over time, these expected benefits have largely materialized, and the development of wood-using industries followed a dependable source of raw materials, which eventually led to further expansion of forest plantations that continues to this date. By the late sixties, when the total plantation area covered approximately 370,000 ha (roughly 20% of the present forest plantation area), plantations had largely surpassed native forests as a source of raw material for the Chilean forest products industry.

Between the early 1930s and early1970s, the Chilean economy benefited from a stable but restrictive policy environment concentrated on internal growth and protectionism. In this policy environment, forest products industries benefited from strong trade protection measures against foreign competition, with effective protection rates way above 100%. Toward the end of this period, and particularly between the mid-sixties and early seventies, plantation-based forest products industries experienced substantial growth.

However, despite all this effective protection, much of this growth was due to an increasing participation of the State in new major facilities (Wisecarver 1988). In other words, the protectionist policy environment was not enough to encourage the private sector to undertake major investments. During this period, drastic land reform policies with a socialistic orientation were introduced. Under the threat of expropriation, private owners were discouraged from committing long-term investments in their properties. As a result, the annual rate of private plantations declined dramatically, and the State began to play a dominant role in forest plantation activities. This brief period of instability in the policy environment was drastically reversed in the mid-seventies.

The policy environment changed drastically with the military takeover of 1973. Basically this was a move away from socialism toward the enhancement of private property rights, a market economy with free pricing mechanisms, an opening to foreign trade and a confinement of the role of the State to activities that would not be properly carried out through private initiative). From a strict forest policy standpoint, new legislation was enacted in 1974 for the purpose of stimulating forest plantations. This new legislation provided for the rebate of 75% of plantation costs up to year 1994. While the majority of the plantations established in the 1970s and 1980s benefited from the rebate, large areas were also planted without the benefit of such incentives. This shows that the confidence in a consistent overall policy environment and the respect for property rights has been a paramount factor in leveraging both the natural economic advantages of forest plantations as well as specific incentives.

Another major factor in the expansion of the Chilean forestry sector, particularly since the mid-seventies has been a major change in policies dealing with trade liberalization. The Chilean economy has become virtually open to world markets in all types of commodities. This opening of foreign markets for all products has allowed Chile to realize its comparative advantage in the forestry sector. For example, exports of all manufactured forest products expanded at an annual rate of more than 12% between the mid-seventies and the mid-eighties, thus enhancing the income of forest plantation owners.

The main lesson to be drawn from the Chilean experience is that private economic agents have created a sizable forest resource. However, the mere presence of a resource is not enough to realize its potential contribution to the wealth of a nation. An overall policy environment allowing for its complete utilization is also necessary. Major changes implemented since the mid-seventies favouring open markets and private property rights have evolved into a stable and predictable policy environment. This has probably been the most important factor favouring the expansion of a diversified forest product economy based on forest plantations.

Privatization of Forest Plantations in New Zealand

From a technical standpoint, New Zealand's experience with forest plantations is not too different from that of Chile. Like in Chile, virtually all the New Zealand forest products economy is based on forest plantations; the main reforestation species, by far, is also Pinus radiata. The forest plantation area is about the same in both countries (1.7 to 1.8 million ha); and productivity and rotation ages are entirely comparable. One minor difference is that plantations in New Zealand started in earnest about 15 to 20 years earlier than in Chile (Scott 1962).

New Zealand has had a long tradition of government-operated enterprises, including postal services, petroleum refining, communications, electrical power, forest production, air transport, marketing of farm products, railroads and others. In the late seventies, however, facing a situation of serious national debt and foreign trade deficits, the major political parties in parliament agreed that many governmental enterprises were being inefficiently operated.

In 1984, 52% of plantation forests in New Zealand were government-owned. This was the result of a long period of dedicated work over which the New Zealand Forest Service achieved outstanding productivity levels, demonstrating a potential imitated on private lands. In the 1980s, however, the Forest Service was accused of being an inefficient organization because of the large number of employees, resulting from continued social policies of subsidizing forest plantation operations to relieve unemployment, provide low cost housing, and keep stumpage prices very low to favour the domestic wood industry (Hall 1997).

The disestablishment of the New Zealand Forest Service began in 1985, and was to be completed in early 1987 with its commercial functions vested in a State-owned enterprise, the New Zealand Forestry Corporation, Ltd. The overall privatization process, however, was not completed until 1996, i.e., 11 years after it was initiated.

The privatization of New Zealand's public forest plantations was a difficult process, which went through several steps and much debate. Initially, there was an internal reorganization in moving from Forest Service to State-owned enterprise, then there were lengthy negotiations on how to set a valuation of the forest estate to be transferred to the newly created enterprise, and it was finally decided to auction the forest assets and let the market determine the values.

The auction process, which started in 1990 and was originally designed to last a few months, was not completed until 1996. The vast majority of public plantations were finally sold, except for a small area under indigenous claims, which is still being managed by State-owned enterprises.

The present status of the New Zealand forestry sector can be summarized as follows:

(i) State ownership of plantations has been reduced from more than 50% in 1989 to less than 5% in 1999;

(ii) over the same period, forest plantation area has expanded from 1.2 to 1.7 million hectares with record plantings from 1994 to 1999;

(iii) major amounts of foreign investment have flowed into plantations and forest products industries, especially from the US, Japan and other Asian sources; and

(iv) forest products exports have increased from NZ$900 million in 1989 to NZ$2.4 billion in 1999. Regarding new plantations, it is worth mentioning that the large majority of these have been conducted by individual owners and small organizations (Koller 2000).

Two major lessons can be derived from the New Zealand experience on privatization of forest plantations. First, although successful in the end, this was a long, complicated and frustrating process, much longer and difficult than originally anticipated. This reflects the importance of the political will of public authorities in seeing the process through. Second, during and after the privatization process, New Zealand's forestry sector has become highly dynamic. While not all the new dynamism can be attributed to the privatization of public plantations, it is clear that the forestry sector now benefits from a much higher participation of private economic agents, both individual and corporate. These new participants have wasted no time in taking advantage of market opportunities, which have been a major factor in the recent expansion of forest plantations.

Tree Plantations and Policy Reforms in China

While the experiences in Chile and New Zealand have shown, albeit in completely different ways, the importance of a combination of open markets and policies as they affect private investors interested in timber production, the analysis of the Chinese experience focuses more on the interactions of forestry and farming in an institutional context in which the influence of State authorities is paramount.

China's central government started focusing on land rehabilitation in the early 1950s. Its initial emphasis was on various engineering and biological measures, including afforestation for windbreaks, for fixing sand, and to control salinization. The very limited new forest resources were rapidly depleted, however, after the collectivization and industrialization movements that began in 1958. Collectivization raised doubts about ownership security for the forest resource and backyard furnaces for steel production consumed much of the remaining wood. The economic adjustments of the early 1960s promoted multi-purpose afforestation activities for environmental protection and improved farm productivity. However, one of the effects of the Cultural Revolution of 1966 and the political struggles that dominated Chinese society for the next decade was that both agricultural and forestry production stagnated.

At the time of its first rural reforms in 1978, China faced a dwindling forest inventory. As a result, it was unable to satisfy the demands of the mills in its wood products industry. The dwindling forest cover also contributed to a major soil erosion problem, decreasing agricultural production, and expanding rural poverty. The first of a series of economy-wide reforms addressed some of the problems in agriculture. The primary reforms gave long-term household responsibility (the Household Responsibility System) for some land that had previously been farmed by the work teams of agricultural collectives, and also allowed households to sell increasing shares of their products in the open market. As a result, grain production increased more than twelve-fold by the mid-1980s.

Forestry gained from these agricultural policy reforms as farmers with newly acquired rights, particularly in the north central region understood the soil conserving properties of trees, and planted trees voluntarily. Harvests of trees increased, but so did forest inventory. Households in more forested regions observed the advantages of household responsibility rights in agriculture and demanded similar long-term rights to forestlands, which they began to obtain in the early 1980s. The first response of many households to the new forest rights was to harvest timber.

Some local forestry authorities reacted by withdrawing the rights from some households and delaying the granting of household responsibility rights to others. The resulting uncertainty of household rights brought about an increase in short-run harvests and the forest inventory declined without replacement. Eventually, however, regions that had endured this policy uncertainty received firm household responsibility rights, leading to increases in both harvests and reinvestments in tree plantations. The response of farmers to household responsibility rights has been most favourable in regions where they have been introduced systematically and much less favourable in areas where their administration has been uneven.

Meanwhile, in the industrial Northeast, State commercial forests predominate, and their mission is to keep the mills supplied with wood. Over time, State administered timber prices have been kept low in order to support the mills. However, low prices have led to insufficient stumpage returns for the Ministry of Forestry. As a result, the Ministry's ability to manage State forests has declined, and so has the forest inventory, and the entire State-run forest industries sector has suffered financial losses since 1990.

The experience of China has demonstrated a number of important lessons. First, even in a highly directed economy, increasingly secure property rights led to the creation of wealth as farmers became aware of the potential benefits of their investments. Second, just like in more market-oriented economies, a stable policy environment has been a major factor in encouraging tree plantations, even if they occur at the modest level of the farm household. Third, farmers, foresters and industrial managers, as economic agents, see clearly the advantages of being able to make decisions in response to market forces. Issues remain however, regarding inefficiencies in State management of industrial forests, and the possibility of an increased role of private initiative in the interest of increased economic efficiency should perhaps be considered.

2

High Value Food Crops

High value crops refer to non-traditional food crops such as vegetables, fruits, flowers, houseplants and foliage, condiments and spices2. High value livestock and fishery products include products such as milk, beef, poultry, pork, eggs, and fish that are non-traditional sources of protein for most part of Sub Saharan Africa (SSA).

Most High Value Agricultural Products (HVAPs) have higher market values than traditional cereal grains and export crops. The definition of HVAP may include coffee and cocoa but these are traditional cash crops in SSA.

A particular feature of HVAPs is that they often do not form part of the customary diet of the local population and are mainly grown for their cash values in domestic and export markets. This review defines HVAP as products with high monetary value with emerging and expanding products markets worldwide.

Lack of a standard definition of HVAP and non-traditional export products has made researchers to adopt their own definitions based on their research emphasis. For example, Xinshen, et. al. (2003) definition of SSA non-traditional HV agricultural exports includes oilseed, processed foods, beverages (coffee, tea and cocoa).

Tea

Tea is an aromatic beverage commonly prepared by pouring hot or boiling water over cured leaves of the tea plant, *Camellia sinensis*. After water, tea is the most widely consumed beverage in the world. It has a cooling, slightly bitter, and astringent flavour that many people enjoy.

Tea originated in China as a medicinal drink. It was first introduced to Portuguese priests and merchants in China during the 16th century.

Drinking tea became popular in Britain during the 17th century. The British introduced it to India, in order to compete with the Chinese monopoly on the product.

Tea has long been promoted for having a variety of positive health benefits. Recent studies suggest that green tea may help reduce the risk of cardiovascular disease and some forms of cancer, promote oral health, reduce blood pressure, help with weight control, improve antibacterial and antivirasic activity, provide protection from solar ultraviolet light, and increase bone mineral density.

Green tea is also said to have "anti-fibrotic properties, and neuroprotective power." Additional research is needed to "fully understand its contributions to human health, and advise its regular consumption in Western diets."

Tea catechins have known anti-inflammatory and neuroprotective properties, help regulate food intake, and have an affinity for cannabinoid receptors, which may suppress pain and nausea and provide calming effects.

Consumption of green tea is associated with a lower risk of diseases that cause functional disability, such as "stroke, cognitive impairment, and osteoporosis" in the elderly.

Tea contains L-theanine, an amino acid whose consumption is mildly associated with a calm but alert and focused, relatively productive (alpha wave-dominant) mental state in humans. This mental state is also common to meditative practice.

The phrase herbal tea usually refers to infusions of fruit or herbs made without the tea plant, such as rosehip tea, chamomile tea, or rooibos tea.

Alternative phrases for this are tisane or herbal infusion, both bearing an implied contrast with "tea" as it is construed here.

Cultivation and Harvesting

Camellia sinensis is an evergreen plant that grows mainly in tropical and subtropical climates.

Some varieties can also tolerate marine climates and are cultivated as far north as Cornwall in the United Kingdom, Washington state in the United States, and experimentally in Pembrokeshire, Wales.

Figure: *Leaves of* Camellia sinensis, *the tea plant*

Tea plants are propagated from seed and cutting; it takes about 4 to 12 years for a tea plant to bear seed and about three years before a new plant is ready for harvesting. In addition to a zone 8 climate or warmer, tea plants require at least 127 cm (50 inches) of rainfall a year and prefer acidic soils. Many high-quality tea plants are cultivated at elevations of up to 1,500 m (4,900 ft) above sea level. While at these heights the plants grow more slowly, they acquire a better flavour.

Two principal varieties are used: *Camellia sinensis* var. *sinensis*, which is used for most Chinese, Formosan and Japanese teas, and *Camellia sinensis* var. *assamica*, used in Pu-erh and most Indian teas (but not Darjeeling). Within these botanical varieties, there are many strains and modern clonal varieties. Leaf size is the chief criterion for the classification of tea plants, with three primary classifications being, Assam type, characterised by the largest leaves; China type, characterised by the smallest leaves; Cambodian type, characterised by leaves of intermediate size.

A tea plant will grow into a tree of up to 16 m (52 ft) if left undisturbed, but cultivated plants are generally pruned to waist height for ease of plucking. Also, the short plants bear more new shoots which provide new and tender leaves and increase the quality of the tea.

Only the top 1–2 inches of the mature plant are picked. These buds and leaves are called *flushes*. A plant will grow a new flush every seven to fifteen days during the growing season. Leaves that are slow in development tend to produce better-flavoured teas. Pests of tea include mosquito bugs that can tatter leaves, so they may be sprayed with insecticides.

Processing and Classification

Teas can generally be divided into categories based on how they are processed. There are at least six different types of tea: white, yellow, green, oolong (or *wulong*), black (called *red tea* in China), and post-fermented tea (or *black tea* for the Chinese) of which the most commonly found on the market are white, green, oolong, and black. Some varieties, such as traditional oolong tea and Pu-erh tea, a post-fermented tea, can be used medicinally.

Figure: *Fresh tea leaves of different sizes: The smaller the leaf, the more expensive the tea.*

After picking, the leaves of *C. sinensis* soon begin to wilt and oxidize, unless they are immediately dried. The leaves turn progressively darker as their chlorophyll breaks down and tannins are released. This enzymatic oxidation process is caused by the plant's intracellular enzymes and causes the tea to darken. In tea processing, the darkening is stopped at a predetermined stage by heating, which deactivates the enzymes responsible. In the production of black teas, the halting of oxidation by heating is carried out simultaneously with drying.

Without careful moisture and temperature control during manufacture and packaging, the tea may become unfit for consumption, due to the growth of undesired molds and bacteria.

Economics

Tea is the most popular manufactured drink in the world in terms of consumption. Its consumption equals all other manufactured drinks in the world – including coffee, chocolate, soft drinks, and alcohol – put together. Most tea consumed outside East Asia is produced on large plantations in the hilly regions of India and Sri Lanka, and is destined to be sold to large businesses. Opposite this large-scale industrial production are many small "gardens," sometimes minuscule plantations, that produce highly sought-after teas prized by gourmets. These teas are both rare and expensive, and can be compared to some of the most expensive wines in this respect.

India is the world's largest tea-drinking nation, although the per capita consumption of tea remains a modest 750 grams per person every year. Turkey, with 2.5 kg of tea consumed per person per year, is the world's greatest per capita consumer.

Production

In 2003, world tea production was 3.21 million tonnes annually. In 2010, world tea production reached over 4.52 million tonnes after having increased by 5.7% between 2009 and 2010. Production rose by 3.1% between 2010 and 2011. The largest producers of tea are the People's Republic of China, India, Kenya, Sri Lanka, and Turkey.

The following table shows the amount of tea production (in tonnes) by leading countries in recent years. Data are generated by the Food and Agriculture Organization of the United Nations as of February 2012.

Certification

Workers who pick and pack tea on plantations in developing countries can face harsh working conditions and can earn below the living wage.

A number of bodies independently certify the production of tea. Tea from certified estates can be sold with a certification label on the pack. The most important certification schemes are Rainforest Alliance, Fairtrade, UTZ Certified, and Organic. All these schemes certify other crops (such as coffee, cocoa and fruit), as well. Rainforest Alliance certified tea is sold by Unilever brands Lipton and PG Tips in Western Europe, Australia and the US. Fairtrade certified tea is sold by a large number of suppliers around the world. UTZ Certified announced a partnership in 2008 with Sara Lee brand Pickwick tea.

Production of organic tea has risen since its introduction in 1990 at Rembeng, Kondoli Tea Estate, Assam. 6,000 tons of organic tea were sold in 1999. About 75% of organic tea production is sold in France, Germany, Japan, the United Kingdom, and the United States.

Trade

According to the FAO in 2007, the largest importer of tea, by weight, was the Russian Federation, followed by the United Kingdom, Pakistan, and the United States. Kenya, China, India and Sri Lanka were the largest exporters of tea in 2007 (with exports of: 374229, 292199, 193459 and 190203 tonnes respectively). The largest exporter of black tea in the world is Kenya, while the largest producer (and consumer) of black tea in the world is India.

Packaging

Tea Bags: In 1907, American tea merchant Thomas Sullivan began distributing samples of his tea in small bags of Chinese silk with a drawstring. Consumers noticed they could simply leave the tea in the bag and reuse it with fresh tea. However, the potential of this distribution/ packaging method would not be fully realised until later on. During World War II, tea was rationed in the United Kingdom. In 1953 (after rationing in the UK ended), Tetley launched the tea bag to the UK and it was an immediate success.

Tea leaves are packed into a small envelope (usually composed of paper) known as a tea bag. The use of tea bags is easy and convenient, making them popular for many people today. However, the use of tea bags has negative aspects, as well. The tea used in tea bags is commonly fannings or "dust", the waste product produced from the sorting of higher-quality loose leaf tea.

However, this is not true for all brands of tea; many high-quality speciality teas are available in bag form. Tea aficionados commonly believe this method provides an inferior taste and experience. The paper used for the bag may also be tasted, which can detract from the tea's own flavour. Because fannings and dust are a lower quality of the tea to begin with, the tea found in tea bags is less finicky when it comes to brewing time and temperature.

Additional reasons why bag tea is considered less well-flavoured include:

- Dried tea loses its flavour quickly on exposure to air. Most bag teas (although not all) contain leaves broken into small pieces; the great surface area to volume ratio of the leaves in tea bags exposes them to more air, and therefore causes them to go stale faster. Loose tea leaves are likely to be in larger pieces, or to be entirely intact.
- Breaking up the leaves for bags extracts flavoured oils.
- The small size of the bag does not allow leaves to diffuse and steep properly.

The "pyramid tea bag" (or sachet) introduced by Lipton and PG Tips/Scottish Blend in 1996, attempts to address one of the connoisseurs' arguments against paper tea bags by way of its three-dimensional tetrahedron shape, which allows more room for tea leaves to expand while steeping. However, some types of pyramid tea bags have been criticised as being environmentally unfriendly, since their synthetic material is not as biodegradable as loose tea leaves and paper tea bags.

Loose Tea

The tea leaves are packaged loosely in a canister, paper bag, or other container such as a tea chest. Some whole teas, such as rolled gunpowder tea leaves, which resist crumbling, are sometimes vacuum packed for freshness in aluminised packaging for storage and retail. The loose tea must be individually measured by the consumer for use in a cup, mug, or teapot.

This allows for flexibility and flavour control, letting the consumer brew weaker or stronger tea as desired, but some convenience is sacrificed compared to newer methods such as tea bags. Strainers, tea balls, tea presses, filtered teapots, and infusion bags can be used to prevent loose leaves from floating in the tea and over-brewing. A more traditional way around this problem is to use a three-piece lidded teacup, called a gaiwan. The lid of the gaiwan can be tilted to decant the leaves while pouring the tea into a different cup for consumption. Alternatively, some tea infusers, sized for a single person and resembling a small metal basket on the end of a spoon's handle, may be used to brew one cup of tea at a time.

Compressed Tea

Some teas (particularly Pu-erh tea) are still compressed for transport, storage, and ageing convenience. The tea brick remains in use in the Himalayan countries or Mongolian steppes. The tea is prepared and steeped by first loosening leaves off the compressed cake using a small knife. Compressed teas can usually be stored for longer periods of time without spoilage when compared with loose leaf tea.

Instant Tea

In recent times, "instant teas" are becoming popular, similar to freeze-dried instant coffee. Similar products also exist for instant iced tea, due to the convenience of not requiring boiling water. Instant tea was developed in the 1930s, but not commercialised until later. Nestlé introduced the first instant tea in 1946, while Redi-Tea introduced the first instant iced tea in 1953.

These products often come with added flavours, such as chai, vanilla, honey or fruit, and may also contain powdered milk. Tea connoisseurs tend to criticise these products for sacrificing the delicacies of tea flavour in exchange for convenience.

British and Canadian soldiers during the Second World War were issued with an instant tea, known as 'Compo' tea. The blocks of instant tea (which included powdered milk and sugar) were issued in a Composite Ration Pack, hence the name 'Compo'. As Royal Canadian Artillery Gunner, George C Blackburn noted, it was not always well received:

But, unquestionably, the feature of Compo rations destined to be remembered beyond all others is Compo tea...Directions say to "sprinkle powder on heated water and bring to the boil, stirring well, three heaped teaspoons to one pint of water."

Every possible variation in the preparation of this tea was tried, but...it always ended up the same way. While still too hot to drink, it is a good-looking cup of strong tea. Even when it becomes just cool enough to be sipped gingerly, it is still a good-tasting cup of tea, if you like your tea strong and sweet. But let it cool enough to be quaffed and enjoyed, and your lips will be coated with a sticky scum that forms across the surface, which if left undisturbed will become a leathery membrane that can be wound around your finger and flipped away...

Bottled and Canned Tea

Switzerland is considered the motherland of bottled iced tea. Maks Sprengler, a Swiss businessman, tried the famous American iced tea and was the first to suggest producing ready-made iced tea in bottles. In 1983, Bischofszell Food Ltd. became the first producer in the world of bottled ice tea on an industrial scale.

Canned tea is a form of tea that has already been prepared, and is sold ready to drink. Canned tea is a fairly recent innovation, first launched in 1981 in Japan.

Storage

Tea shelf life varies with storage conditions and type of tea. Black tea has a longer shelf life than green tea. An exception, pu-erh tea, improves with age. Tea stays freshest when stored in a dry, cool, dark place in an air-tight container. Black tea stored in a bag inside a sealed opaque canister may keep for two years. Green tea loses its freshness more quickly, usually in less than a year. Gunpowder tea, its leaves being tightly rolled, keeps longer than the more open-leafed Chun Mee tea. Storage life for all teas can be extended by using desiccant packets, oxygen-absorbing packets, vacuum sealing or store tea in closed containers in a refrigerator.

When storing green tea, discreet use of refrigeration or freezing is recommended. In particular, drinkers need to take precautions against temperature variation. Improperly stored tea may lose flavour, acquire disagreeable flavours or odors from other foods, or become moldy.

Coffee

Coffee is a brewed beverage prepared from the roasted or baked seeds of several species of an evergreen shrub of the genus *Coffea*. The two most common sources of coffee beans are the highly regarded *Coffea arabica*, and the "robusta" form of the hardier *Coffea canephora*. The latter is resistant to the coffee leaf rust (*Hemileia vastatrix*), but has a more bitter taste. Coffee plants are cultivated in more than more than 70 countries, primarily in equatorial Latin America, Southeast Asia, and Africa.

Once ripe, coffee "berries" are picked, processed and dried to yield the seeds inside. The seeds are then roasted to varying degrees, depending on the desired flavour, before being ground and brewed to create coffee.

Coffee is slightly acidic (pH 5.0–5.1) and can have a stimulating effect on humans because of its caffeine content. It is one of the most popular drinks in the world. It can be prepared and presented in a variety of ways.

The effect of coffee on human health has been a subject of many studies; however, results have varied in terms of coffee's relative benefit. The majority of recent research suggests that moderate coffee consumption is benign or mildly beneficial in healthy adults. However, the diterpenes in coffee may increase the risk of heart disease.

Coffee cultivation first took place in southern Arabia; the earliest credible evidence of coffee-drinking appears in the middle of the 15th century in the Sufi shrines of Yemen.

In East Africa and Yemen, coffee was used in native religious ceremonies. As these ceremonies conflicted with the beliefs of the Christian church, the Ethiopian Church banned the secular consumption of coffee until the reign of Emperor Menelik II. The beverage was also banned in Ottoman Turkey during the 17th century for political reasons and was associated with rebellious political activities in Europe.

An important export commodity, coffee was the top agricultural export for twelve countries in 2004, and it was the world's seventh-largest legal agricultural export by value in 2005.

Green (unroasted) coffee is one of the most traded agricultural commodities in the world. Some controversy is associated with coffee cultivation and its impact on the environment. Consequently, organic coffee is an expanding market.

Biology

Figure: *Illustration of Coffea arabica plant and seeds*

Several species of shrub of the genus *Coffea* produce the berries from which coffee is extracted. The two main species commercially cultivated are *Coffea canephora* (predominantly a form known as 'robusta') and *C. arabica*. *C. arabica*, the most highly regarded species, is native to the southwestern highlands of Ethiopia and the Boma Plateau in southeastern Sudan and possibly Mount Marsabit in northern Kenya. *C. canephora* is native to western and central Subsaharan Africa, from Guinea to the Uganda and southern Sudan. Less popular species are *C. liberica*, *C. stenophylla*, *C. mauritiana*, and *C. racemosa*.

All coffee plants are classified in the large family Rubiaceae. They are evergreen shrubs or small trees that may grow 5 m (15 ft) tall when unpruned. The leaves are dark green and glossy, usually 10–15 cm (4–6 in) long and 6 cm (2.4 in) wide, simple, entire, and opposite. Petioles of opposite leaves fuse at base to form interpetiolar stipules, characteristic of Rubiaceae. The flowers are axillary, and clusters of fragrant white flowers bloom simultaneously. Gynoecium consists of inferior ovary, also characteristic of Rubiaceae. Flowers followed by oval berries of about 1.5 cm (0.6 in). Green when immature, they ripen to yellow, then crimson, before turning black on drying. Each berry usually contains two seeds, but 5–10% of the berries have only one; these are called peaberries. Arabica berries ripen in six to eight months, while robusta take nine to eleven months.

Coffea arabica is predominantly self-pollinating, and as a result the seedlings are generally uniform and vary little from their parents. In contrast, *Coffea canephora*, and *C. liberica* are self-incompatible and require outcrossing. This means that useful forms and hybrids must be propagated vegetatively. Cuttings, grafting, and budding are the usual methods of vegetative propagation. On the other hand, there is great scope for experimentation in search of potential new strains.

Cultivation

The traditional method of planting coffee is to place 20 seeds in each hole at the beginning of the rainy season. This method loses about 50% of the seeds' potential, as about half fail to sprout. A more effective method of growing coffee, used in Brazil, is to raise seedlings in nurseries that are then planted outside at six to twelve months. Coffee is often intercropped with food crops, such as corn, beans, or rice during the first few years of cultivation as farmers become familiar with its requirements.

Of the two main species grown, arabica coffee (from *C. arabica*) is generally more highly regarded than robusta coffee (from *C. canephora*); robusta tends to be bitter and have less flavour but better body than arabica. For these reasons, about three-quarters of coffee cultivated worldwide is *C. arabica*. Robusta strains also contain about 40–50% more caffeine than arabica. For this reason, it is used as an inexpensive substitute for arabica in many commercial coffee blends. Good quality robusta seeds are used in traditional Italian espresso blends to provide a full-bodied taste and a better foam head (known as *crema*).

However, *Coffea canephora* is less susceptible to disease than *C. arabica* and can be cultivated in lower altitudes and warmer climates where *C. arabica* will not thrive. The robusta strain was first collected in 1890 from the Lomani River, a tributary of the Congo River, and was conveyed from Zaire (now the Democratic Republic of Congo) to Brussels to Java around 1900.

From Java, further breeding resulted in the establishment of robusta plantations in many countries. In particular, the spread of the devastating coffee leaf rust (*Hemileia vastatrix*), to which *C. arabica* is vulnerable, hastened the uptake of the resistant robusta. Coffee leaf rust is found in virtually all countries that produce coffee.

Over 900 species of insect have been recorded as pests of coffee crops worldwide. Of these, over a third are beetles, and over a quarter are bugs. Some 20 species of nematodes, 9 species of mites, several snails and slugs also attack the crop. Birds and rodents sometimes eat coffee berries but their impact is minor compared to invertebrates. In general, *arabica* is the more sensitive species to invertebrate predation overall. Each part of the coffee plant is assailed by different animals. Nematodes attack the roots, and borer beetles burrow into stems and woody material, the foliage is attacked by over 100 species of larvae (caterpillars) of butterflies and moths.

Mass spraying of insecticides has often proven disastrous, as the predators of the pests are more sensitive than the pests themselves. Instead, integrated pest management has developed, using techniques such as targeted treatment of pest outbreaks, and managing crop environment away from conditions favouring pests. Branches infested with scale are often cut and left on the ground, which promotes scale parasites to not only attack the scale on the fallen branches but in the plant as well.

The 2-mm-long coffee berry borer beetle is the most damaging insect pest to the world's coffee industry, destroying up to 50 percent or more of the coffee berries on plantations in most coffee-producing countries. The adult female beetle nibbles a single tiny hole in a coffee berry and lays 35 to 50 eggs.

Inside, the offspring grow, mate, and then emerge from the commercially ruined berry to disperse, repeating the cycle. Pesticides are mostly ineffective because the beetle juveniles are protected inside the berry nurseries, but they are vulnerable to predation by birds when they emerge.

When groves of trees are nearby, the American Yellow Warbler, Rufous-capped Warbler and other insectivorous birds have been shown to reduce by 50 percent the number of coffee berry borer beetles in Costa Rica coffee plantations.

World Production

In 2011 Brazil was the world leader in production of green coffee, followed by Vietnam, Indonesia and Colombia. Arabica coffee seeds are cultivated in Latin America, eastern Africa, Arabia, or Asia. Robusta coffee seeds are grown in western and central Africa, throughout southeast Asia, and to some extent in Brazil.

Seeds from different countries or regions can usually be distinguished by differences in flavour, aroma, body, and acidity. These taste characteristics are dependent not only on the coffee's growing region, but also on genetic subspecies (varietals) and processing. Varietals are generally known by the region in which they are grown, such as Colombian, Java and Kona.

Ecological Effects

Originally, coffee farming was done in the shade of trees that provided a habitat for many animals and insects. Remnant forest trees were used for this purpose, but many species have been planted as well. These include leguminous trees of the genera *Acacia*, *Albizia*, *Cassia*, *Erythrina*, *Gliricidia*, *Inga*, and *Leucaena*, as well as the nitrogen-fixing non-legume sheoaks of the genus *Casuarina*, and the silky oak *Grevillea robusta*.

This method commonly referred to as the traditional shaded method, or "shade-grown". Starting in the 1970s, many farmers switched their production method to sun cultivation, in which coffee is grown in rows under full sun with little or no forest canopy. This causes berries to ripen more rapidly and bushes to produce higher yields, but requires the clearing of trees and increased use of fertilizer and pesticides, which damage the environment and cause health problems.

Unshaded coffee plants grown with fertilizer yield the most coffee, although unfertilized shaded crops generally yield more than unfertilized unshaded crops: the response to fertilizer is much greater in full sun. Although traditional coffee production causes berries to ripen more slowly and produce lower yields, the quality of the coffee is allegedly superior. In addition, the traditional shaded method provides living space for many wildlife species. Proponents of shade cultivation say environmental problems such as deforestation, pesticide pollution, habitat destruction, and soil and water degradation are the side effects of the practices employed in sun cultivation.

The American Birding Association, Smithsonian Migratory Bird Centre, National Arbor Day Foundation, and the Rainforest Alliance have led a campaign for 'shade-grown' and organic coffees, which can be sustainably harvested. Shaded coffee cultivation systems show greater biodiversity than full-sun systems, and those more distant from continuous forest compare rather poorly to undisturbed native forest in terms of habitat value for some bird species.

Another issue concerning coffee is its use of water. According to *New Scientist*, using industrial farming practices, it takes about 140 litres (37 U.S. gal) of water to grow the coffee seeds needed to produce one cup of coffee, and the coffee is often grown in countries where there is a water shortage, such as Ethiopia.

By using sustainable agriculture methods, the amount of water usage can be dramatically reduced, while retaining comparable yields. For comparison, the United States Geological Survey reports that one egg requires an input of 454 litres (120 U.S. gal) of water; one serving of milk requires an input of 246 litres (65 U.S. gal) of water; one serving of rice requires an input of 132 litres (35 U.S. gal) of water; and one glass of wine requires an input of 120 litres (32 U.S. gal) of water.

Coffee grounds may be used for composting or as a mulch. They are especially appreciated by worms and acid-loving plants such as blueberries. Some commercial coffee shops run initiatives to make better use of these grounds, including Starbucks' "Grounds for your Garden" project, and community sponsored initiatives such as "Ground to Ground".

Starbucks sustainability chief Jim Hanna has warned that Climate change may significantly impact coffee yields within a few decades.

Production

Figure: *Ripe coffee berries*

Figure: *Unripe coffee berries near Kona, Hawaii*

Processing

Coffee berries and their seeds undergo several processes before they become the familiar roasted coffee. Berries have been traditionally selectively picked by hand; a labour-intensive method, it involves the selection of only the berries at the peak of ripeness. More commonly, crops are strip picked, where all berries are harvested simultaneously regardless of ripeness by person or machine. After picking, green coffee is processed by one of two methods—the *dry process method*, simpler and less labour-intensive as the berries can be strip picked, and the *wet process method*, which incorporates fermentation into the process and yields a mild coffee.

Then they are sorted by ripeness and colour and most often the flesh of the berry is removed, usually by machine, and the seeds are fermented to remove the slimy layer of mucilage still present on the seed. When the fermentation is finished, the seeds are washed with large quantities of fresh water to remove the fermentation residue, which generates massive amounts of coffee wastewater. Finally, the seeds are dried.

The best (but least used) method of drying coffee is using drying tables. In this method, the pulped and fermented coffee is spread thinly on raised beds, which allows the air to pass on all sides of the coffee, and then the coffee is mixed by hand. In this method the drying that takes place is more uniform, and fermentation is less likely. Most African coffee is dried in this manner and certain coffee farms around the world are starting to use this traditional method.

Next, the coffee is sorted, and labelled as green coffee. Another way to let the coffee seeds dry is to let them sit on a concrete patio and rake over them in the sunlight. Some companies use cylinders to pump in heated air to dry the coffee seeds, though this is generally in places where the humidity is very high.

Some coffee undergoes a peculiar process, such as kopi luwak. It is made from the seeds of coffee berries which have been eaten by the Asian Palm Civet and other related civets, passing through its digestive tract. This process resulted in coffee seeds with much less bitterness, widely noted as the most expensive coffee in the world with prices reaching $160 per pound.

Roasting

The next step in the process is the roasting of the green coffee. Coffee is usually sold in a roasted state, and with rare exceptions all coffee is roasted before it is consumed. It can be sold roasted by the supplier, or it can be home roasted. The roasting process influences the taste of the beverage by changing the coffee seed both physically and chemically.

The seed decreases in weight as moisture is lost and increases in volume, causing it to become less dense. The density of the seed also influences the strength of the coffee and requirements for packaging.

The actual roasting begins when the temperature inside the seed reaches approximately 200 °C (392 °F), though different varieties of seeds differ in moisture and density and therefore roast at different rates. During roasting, caramelization occurs as intense heat breaks down starches, changing them to simple sugars that begin to brown, which alters the colour of the seed.

Sucrose is rapidly lost during the roasting process and may disappear entirely in darker roasts. During roasting, aromatic oils and acids weaken, changing the flavour; at 205 °C (401 °F), other oils start to develop. One of these oils, caffeol, is created at about 200 °C (392 °F), which is largely responsible for coffee's aroma and flavour.

Grading the Roasted Seeds

Depending on the colour of the roasted seeds as perceived by the human eye, they will be labelled as light, medium light, medium, medium dark, dark, or very dark. A more accurate method of discerning the degree of roast involves measuring the reflected light from roasted seeds illuminated with a light source in the near infrared spectrum. This elabourate light metre uses a process known as spectroscopy to return a number that consistently indicates the roasted coffee's relative degree of roast or flavour development.

Roast Characteristics

The degree of roast has an effect upon coffee flavour and body. Darker roasts are generally bolder because they have less fibre content and a more sugary flavour. Lighter roasts have a more complex and therefore perceived stronger flavour from aromatic oils and acids otherwise destroyed by longer roasting times. Roasting does not alter the amount of caffeine in the bean, but does give less caffeine when the beans are measured by volume because the beans expand during roasting.

A small amount of chaff is produced during roasting from the skin left on the seed after processing. Chaff is usually removed from the seeds by air movement, though a small amount is added to dark roast coffees to soak up oils on the seeds.

Decaffeination

Decaffeination may also be part of the processing that coffee seeds undergo. Seeds are decaffeinated when they are still green. Many methods can remove caffeine from coffee, but all involve either soaking the green seeds in hot water (often called the "Swiss water process") or steaming them, then using a solvent to dissolve caffeine-containing oils. Decaffeination is often done by processing companies, and the extracted caffeine is usually sold to the pharmaceutical industry.

Storage

Coffee is best stored in an airtight container made of ceramic, glass, or non-reactive metal. Higher quality prepackaged coffee usually has a one-way valve which prevents air from entering while allowing the coffee to release gases. Coffee freshness and flavour is preserved when it is stored away from moisture, heat, and light. The ability of coffee to absorb strong smells from food means that it should be kept away from such smells. Storage of coffee in the refrigerator is not recommended due to the presence of moisture which can cause deterioration. Exterior walls of buildings which face the sun may heat the interior of a home, and this heat may damage coffee stored near such a wall. Heat from nearby ovens also harms stored coffee.

In 1931, a method of packing coffee in a sealed vacuum in cans was introduced. The roasted coffee was packed and then 99% of the air was removed, allowing the coffee to be stored indefinitely until the can was opened. Today this method is in mass use for coffee in a large part of the world.

Brewing

Coffee seeds must be ground and brewed to create a beverage. The criteria for choosing a method include flavour and economy. Almost all methods of preparing coffee require that the seeds be ground and then mixed with hot water long enough to allow the flavour to emerge but not so long as to draw out bitter compounds. The liquid can be consumed after the spent grounds are removed. Brewing considerations include the fineness of grind, the way in which the water is to extract the flavour, additional flavorings such as sugar, milk, and spices, and the technique to be used to separate spent grounds. Ideal holding temperatures range from 85–88 °C (185–190 °F) to as high as 93 °C (199 °F) and the ideal serving temperature is 68 to 79 °C (154 to 174 °F).

The roasted coffee seeds may be ground at a roastery, in a grocery store, or in the home. Most coffee is roasted and ground at a roastery and sold in packaged form, though roasted coffee seeds can be ground at home immediately before consumption. It is also possible, though uncommon, to roast raw seeds at home.

The choice of brewing method depends to some extent on the degree to which the coffee seeds have been roasted. Lighter roasted coffee tends to be used for filter coffee as the combination of method and roast style results in higher acidity, complexity, and clearer nuances. Darker roasted coffee is used for espresso because the machine naturally extracts more dissolved solids, causing lighter coffee to become too acidic.

Coffee seeds may be ground in several ways. A burr grinder uses revolving elements to shear the seed; a blade grinder cuts the seeds with blades moving at high speed; and a mortar and pestle crushes the seeds. For most brewing methods a burr grinder is deemed superior because the grind is more even and the grind size can be adjusted.

The type of grind is often named after the brewing method for which it is generally used. Turkish grind is the finest grind, while coffee percolator or French press are the coarsest grinds. The most common grinds are between these two extremes: a medium grind is used in most home coffee-brewing machines.

Coffee may be brewed by several methods. It may be boiled, steeped, or pressurized.

Brewing coffee by boiling was the earliest method, and Turkish coffee is an example of this method. It is prepared by grinding or pounding the seeds to a fine powder, then adding it to water and bringing it to the boil for no more than an instant in a pot called a *cezve* or, in Greek, a *bríki*. This produces a strong coffee with a layer of foam on the surface and sediment (which is not meant for drinking) settling at the bottom of the cup.

Coffee percolators and automatic coffeemakers brew coffee using gravity. In an automatic coffeemaker, hot water drips onto coffee grounds that are held in a paper, plastic, or perforated metal coffee filter, allowing the water to seep through the ground coffee while extracting its oils and essences. The liquid drips through the coffee and the filter into a carafe or pot, and the spent grounds are retained in the filter.

In a percolator, boiling water is forced into a chamber above a filter by steam pressure created by boiling. The water then seeps through the grounds, and the process is repeated until terminated by removing from the heat, by an internal timer, or by a thermostat that turns off the heater when the entire pot reaches a certain temperature.

Coffee may be brewed by steeping in a device such as a French press (also known as a *cafetière*, coffee press or coffee plunger). Ground coffee and hot water are combined in a cylindrical vessel and left to brew for a few minutes. A circular filter which fits tightly in the cylinder fixed to a plunger is then pushed down from the top to force the grounds to the bottom. The filter retains the grounds at the bottom as the coffee is poured from the container. Because the coffee grounds are in direct contact with the water, all the coffee oils remain in the liquid, making it a stronger beverage.

This method of brewing leaves more sediment than in coffee made by an automatic coffee machine. Supporters of the French press method point out that the sediment issue can be minimized by using the right type of grinder: they claim that a rotary blade grinder cuts the coffee bean into a wide range of sizes, including a fine coffee dust that remains as sludge at the bottom of the cup, while a burr grinder uniformly grinds the beans into consistently-sized grinds, allowing the coffee to settle uniformly and be trapped by the press. 95% of the caffeine is released from the coffee seeds within the first minute of brewing.

The espresso method forces hot pressurized and vaporized water through ground coffee. As a result of brewing under high pressure (ideally between 9–10 atm), the espresso beverage is more concentrated (as much as 10 to 15 times the quantity of coffee to water as gravity-brewing methods can produce) and has a more complex physical and chemical constitution. A well-prepared espresso has a reddish-brown foam called *crema* that floats on the surface. Other pressurized water methods include the moka pot and vacuum coffee maker.

Cold brew coffee is made by steeping coarsely ground seeds in cold water for several hours, then filtering them. This results in a brew lower in acidity than most hot-brewing methods.

Serving

Once brewed, coffee may be served in a variety of ways. Drip-brewed, percolated, or French-pressed/cafetière coffee may be served as *white coffee* with a dairy product such as milk or cream, or dairy substitute, or as *black coffee* with no such addition. It may be sweetened with sugar or artificial sweetener. When served cold, it is called *iced coffee*.

Espresso-based coffee has a variety of possible presentations. In its most basic form, an espresso is served alone as a *shot* or *short black*, or with hot water added, and is known as Caffè Americano. A long black is made by pouring a double espresso into an equal portion of water, retaining the crema, unlike Caffè Americano. Milk is added in various forms to an espresso: steamed milk makes a caffè latte, equal parts steamed milk and milk froth make a cappuccino, and a dollop of hot foamed milk on top creates a caffè macchiato. The use of steamed milk to form patterns such as hearts or maple leaves is referred to as latte art.

Coffee can also be incorporated with alcohol to produce a variety of beverages: it is combined with whiskey in Irish coffee, and it forms the base of alcoholic coffee liqueurs such as Kahlúa and Tia Maria. Coffee is also sometimes used in the brewing process of darker beers, such as a stout or porter.

Instant Coffee

A number of products are sold for the convenience of consumers who do not want to prepare their own coffee.

Instant coffee is dried into soluble powder or freeze-dried into granules that can be quickly dissolved in hot water. Originally invented in 1907, it rapidly gained in popularity in many countries in the post-war period, with Nescafé being the most popular product. Many consumers determined that the convenience in preparing a cup of instant coffee more than made up for a perceived inferior taste. Paralleling (and complementing) the rapid rise of instant coffee was the coffee vending machine, invented in 1947 and multiplying rapidly through the 1950s.

Canned coffee has been popular in Asian countries for many years, particularly in China, Japan, South Korea, and Taiwan. Vending machines typically sell varieties of flavored canned coffee, much like brewed or percolated coffee, available both hot and cold. Japanese convenience stores and groceries also have a wide availability of bottled coffee drinks, which are typically lightly sweetened and pre-blended with milk. Bottled coffee drinks are also consumed in the United States.

Liquid coffee concentrates are sometimes used in large institutional situations where coffee needs to be produced for thousands of people at the same time. It is described as having a flavour about as good as low-grade robusta coffee, and costs about 10¢ a cup to produce. The machines can process up to 500 cups an hour, or 1,000 if the water is preheated.

Competition

Coffee competitions take place across the globe with people at the regional competing to achieve national titles and then compete on the international stage. World Coffee Events holds the largest of such events moving the location of the final competition each year. The competition includes the following events: Barista Championship,

Brewers Cup, Latte Art and Cup Tasters. A World Brewer's Cup Championship takes place in Melbourne, Australia every year that houses contestants from around the world to crown the World's Coffee King.

Indigofera

Indigofera is a large genus of over 750 species of flowering plants belonging to the family Fabaceae. They are widely distributed throughout the tropical and subtropical regions of the world.

Figure: *Indigofera tinctoria*

Description

Species of *Indigofera* are mostly shrubs, though some are small trees or annual or perennial herbs. Most have pinnate leaves. Racemes of flowers grow in the leaf axils. Most species have flowers in shades of red, but there are a few white- and yellow-flowered species. The fruit is a legume pod of varying size and shape.

Uses

Indigo Dye: Several species, especially *Indigofera tinctoria* and *Indigofera suffruticosa*, are used to produce the dye indigo. Colonial planters in the Caribbean grew indigo and transported its cultivation when they settled in the colony of South Carolina and North Carolina Where people of the Tuscarora confederacy adopted the dying process for head wraps and clothing. Exports of the crop did not expand until the mid-to late 18th century. When Eliza Lucas Pinckney and enslaved

Africans successfully cultivated new strains near Charleston it became the second most important cash crop in the colony (after rice) before the American Revolution. It comprised more than one-third of all exports in value.

The chemical aniline, from which many important dyes are derived, was first synthesized from *I. suffruticosa* (syn. *I. anil*, whence the name aniline).

In Indonesia, the Sundanese use *Indigofera tinctoria* (known locally as *tarum*) as dye for batik.

Health and Medicine

Several species of this group are used to alleviate pain. The herbs are generally regarded as an analgesic with anti-inflammatory activity, rather than an anodyne. *Indigofera articulata* (*Khedaish* in Arabic) was used for toothache, and *Indigofera oblongifolia* (*hasr* in Arabic) was used as an anti-inflammatory for insect stings, snakebites, and swellings.

Indigofera suffruticosa and *Indigofera aspalthoides* have also been used as anti-inflammatories. A patent was granted for use of *Indigofera arrecta* extract to relieve ulcer pain.

The Maasai people of Kenya use parts of *Indigofera brevicalyx* and *I. swaziensis* as toothbrushes.

Ecology

Indigofera species are used as food plants by the larvae of some Lepidoptera species, including the Turnip Moth (*Agrotis segetum*).

Cacao

Theobroma cacao also cacao tree and cocoa tree, is a small (4–8 m (13–26 ft) tall) evergreen tree in the family Malvaceae, native to the deep tropical regions of Central and South America. Its seeds, cocoa beans, are used to make cocoa mass, cocoa powder, and chocolate.

Leaves are alternate, entire, unlobed, 10–40 cm (3.9–15.7 in) long and 5–20 cm (2.0–7.9 in) broad.

The flowers are produced in clusters directly on the trunk and older branches; this is known as cauliflory. The flowers are small, 1–2 cm (0.39–0.79 in) diametre, with pink calyx. While many of the world's flowers are pollinated by bees (Hymenoptera) or butterflies/moths (Lepidoptera), cacao flowers are pollinated by tiny flies, *Forcipomyia* midges in the order Diptera. The fruit, called a cacao pod, is ovoid, 15–30 cm (5.9–11.8 in) long and 8–10 cm (3.1–3.9 in) wide, ripening yellow to orange, and weighs about 500 g (1.1 lb) when ripe. The pod contains

20 to 60 seeds, usually called "beans", embedded in a white pulp. The seeds are the main ingredient of chocolate, while the pulp is used in some countries to prepare refreshing juice, smoothies, jelly, and nata. Each seed contains a significant amount of fat (40–50%) as cocoa butter. Their most noted active constituent is theobromine, a compound similar to caffeine.

Taxonomy and Nomenclature

Cacao (*Theobroma cacao*) belongs to the genus *Theobroma* classified under the subfamily Sterculioidea of the mallow family Malvaceae. Cacao is one of 22 species of *Theobroma*.

The specific name *cacao* is derived from the native name of the plant in indigenous Mesoamerican languages. The cacao was known as *kakaw* in Tzeltal, K'iche' and Classic Maya; *kagaw* in Sayula Popoluca; and *cacahuatl* in Nahuatl.

Cupuaçu, *Theobroma grandiflorum*, is a closely related species found in Colombia, Peru, Bolivia and Brazil. Like cacao, it is also the source for a kind of chocolate known as *cupulate* or cupuaçu chocolate. Cupuaçu is considered as having high potential by the food and cosmetics industries.

Distribution and Domestication

T. cacao is widely distributed from southeastern Mexico to the Amazon basin. There were originally two hypotheses about its domestication; one said that there were two foci for domestication, one in the Lacandon area of Mexico and another in lowland South America. More recent studies of patterns of DNA diversity, however, suggest that this is not the case. Motomayor *et al.* sampled 1241 trees and classified them into 10 distinct genetic clusters. This study also identified areas, for example around Iquitos in modern Peru, where representatives of several genetic clusters originated. This result suggests that this is where *T. cacao* was originally domesticated, probably for the pulp that surrounds the beans, which is eaten as a snack and fermented into a mildly alcoholic beverage.

Using the DNA sequences obtained by Motomayor *et al.* and comparing them with data derived from climate models and the known conditions suitable for cacao, Thomas *et al.* have further refined the view of domestication, linking the area of greatest cacao genetic diversity to a bean-shaped area that encompasses the border between Brazil and Peru and the southern part of the Colombian-Brazilian border. Climate models indicate that at the peak of the last ice age 21,000 years ago, when habitat suitable for cacao was at its most reduced, this area was still suitable, and so provided a refugium for the species. Thomas *et al.* speculate that from there people took cacao to Mexico, where selection for the beans took place.

Cacao trees grow well as understory plants in humid forest ecosystems. This is equally true of abandoned cultivated trees, making it difficult to distinguish truly wild trees from those whose parents may originally have been cultivated.

History of Cultivation

Cultivation, use, and cultural elaboration of cacao were early and extensive in Mesoamerica. Ceramic vessels with residues from the preparation of cacao beverages have been found at archaeological sites dating back to the Early Formative (1900-900 BC) period. For example, one such vessel found at an Olmec archaeological site on the Gulf Coast of Veracruz, Mexico dates cacao's preparation by pre-Olmec peoples as early as 1750 BC. On the Pacific coast of Chiapas, Mexico, a Mokaya archaeological site provides evidence of cacao beverages dating even earlier, to 1900 BC. The initial domestication was probably related to the making of a fermented, thus alcoholic beverage.

Several mixtures of cacao are described in ancient texts, for ceremonial or medicinal, as well as culinary, purposes. Some mixtures included maize, chili, vanilla (*Vanilla planifolia*), and honey. Archaeological evidence for use of cacao, while relatively sparse, has come from the recovery of whole cacao beans at Uaxactun, Guatemala and from the preservation of wood fragments of the cacao tree at Belize sites including Cuello and Pulltrouser Swamp.

In addition, analysis of residues from ceramic vessels has found traces of theobromine and caffeine in early formative vessels from Puerto Escondido, Honduras (1100-900 BC) and in middle formative vessels from Colha, Belize (600-400 BC) using similar techniques to those used to extract chocolate residues from four classic period (*circa* 400 AD) vessels from a tomb at the Maya archaeological site of Rio Azul. As cacao is the only known commodity from Mesoamerica containing both of these alkaloid compounds, it seems likely these vessels were used as containers for cacao drinks. In addition, cacao is named in a hieroglyphic text on one of the Rio Azul vessels. Cacao was also believed to be ground by the Aztecs and mixed with tobacco for smoking purposes.

Currency System

Cacao beans constituted both a ritual beverage and a major currency system in pre-Columbian Mesoamerican civilizations. At one point, the Aztec empire received a yearly tribute of 980 loads (*xiquipil* in Nahuatl) of cacao, in addition to other goods. Each load represented exactly 8,000 beans. The buying power of quality beans was such that 80-100 beans could buy a new cloth mantle. The use of cacao beans as currency is also known to have spawned counterfeiters during the Aztec empire.

Cultivation

Cacao is cultivated on roughly 17,000,000 acres (27,000 sq mi; 69,000 km^2) worldwide. According to the Food and Agriculture Organization (FAO), the top 20 cacao-producing countries in 2005 were as follows:

Cacao production has increased from 1.5 million tons in 1983-1984 to 3.5 million tons in 2003-2004, almost entirely due to the expansion of the production area rather than to yield increases. Cacao is grown both by large agroindustrial plantations and small producers, the bulk of production coming from millions of farmers who have a few trees each.

A tree begins to bear when it is four or five years old. A mature tree may have 6,000 flowers in a year, yet only about 20 pods. About 1,200 seeds (40 pods) are required to produce 1 kg (2.2 lb) of cocoa paste.

Historically, chocolate makers have recognized three main cultivar groups of cacao beans used to make cocoa and chocolate. The most prized, rare, and expensive is the Criollo group, the cocoa bean used by the Maya. Only 10% of chocolate is made from Criollo, which is less bitter and more aromatic than any other bean. The cacao bean in 80% of chocolate is made using beans of the Forastero group. Forastero trees are significantly hardier than Criollo trees, resulting in cheaper cacao beans. Trinitario, a hybrid of Criollo and Forastero, is used in about 10% of chocolate. The criollo cacao beans from Chuao in Aragua, Venezuela are widely regarded as some of the finest in the world. In November 2000, the cacao beans coming from said region were awarded an appellation of origin under the title "Cacao de Chuao"(from Spanish-*cacao of Chuao*) effectively making this one of the most expensive and sought after types of cacao.

A new, genetically based classification of 10 groups may well help breeders to create new varieties that are both pest- and disease-resistant and contain valued flavours.

Major cocoa bean processors include Hershey's, Nestlé and Mars, all of which purchase cocoa beans via various sources.

In June 2009, Mars Botanicals, a division of Mars, launched Cirku, a cocoa extract product that provides cocoa ûavanols made with a patented process that contains a high level of phytonutrients.

Pests

Various plant pests and diseases can cause serious problems for cacao production.

- Insects
  - Cocoa mirids or capsids worldwide (but especially *Sahlbergella singularis* and *Distantiella theobroma* in West Africa and *Helopeltis* spp. in Southeast Asia)
  - *Conopomorpha cramerella* (cocoa pod borer - in Southeast Asia)
- Fungi
  - *Moniliophthora roreri* (frosty pod rot)
  - *Moniliophthora perniciosa* (witches' broom)
  - *Ceratocystis cacaofunesta* (*mal de machete*) or (*Ceratocystis* wilt)
  - *Verticillium dahliae*
  - *Oncobasidium theobromae* (vascular streak dieback)
- Oomycetes
  - *Phytophthora* spp. (black pod) especially *Phytophthora megakarya* in West Africa
- Viruses
  - CSSV
- Mistletoe
- Rats and other vertebrate pests (squirrels, woodpeckers, *etc.*)

Conservation

The pests and diseases to which cacao is subject, along with climate change, mean that new varieties will be needed to respond to these challenges. Breeders rely on the genetic diversity conserved in field genebanks to create new varieties, because cacao has recalcitrant seeds that cannot be stored in a conventional genebank. In an effort to improve the diversity available to breeders, and ensure the future of the field genebanks, experts have drawn up A Global Strategy for the Conservation and Use of Cacao Genetic Resources, as the Foundation for a Sustainable Cocoa Economy. The strategy has been adopted by the cacao producers and their clients, and seeks to improve the characterization of cacao diversity, the sustainability and diversity of the cacao collections, the usefulness of the collections, and to ease access to better information about the conserved material. Some natural areas of cacao diversity are protected by various forms of conservation, for example national parks.

However, a recent study of genetic diversity and predicted climates suggests that many of those protected areas will no longer be suitable for cacao by 2050. It also identifies an area around Iquitos in Peru that will remain suitable for cacao and that is home to considerable genetic diversity, and recommends that this area be considered for protection.

Cacao Genome

The genome of *T. cacao* is diploid, its size is 430 Mbp, and it comprises 10 chromosome pairs (2n=2x=20). In September 2010, a team of scientists announced a draft sequence of the cacao genome (Matina1-6 genotype).

In a second, unrelated project, the International Cocoa Genome Sequencing Consortium-ICGS, co-ordinated by CIRAD, first published in December 2010 (online, paper publication in January 2011), the sequence of the cacao genome, of the Criollo cacao (of a landrace from Belize, B97-61/B2). In their publication, they reported a detailed analysis of the genomic and genetic data.

The sequence of the cacao genome identified 28,798 protein-coding genes, compared to the roughly 23,000 protein-coding genes of the human genome. About 20% of the cacao genome consists of transposable elements, a low proportion compared to other plant species. Many genes were identified as coding for flavonoids, aromatic terpenes, theobromine and many other metabolites involved in cocoa flavour and quality traits, among which a relatively high proportion code for polyphenols, which constitute up to 8% of cacao pods dry weight.

The cacao genome appears close to the hypothetical hexaploid ancestor of all dicotyledonous plants, and it is proposed as an evolutionary mechanism by which the 21 chromosomes of the dicots' hypothetical hexaploid ancestor underwent major fusions leading to cacao's 10 chromosome pairs.

The genome sequence will help accelerate research on cacao molecular biology and breeding for elite varieties through marker-assisted selection, in particular for genetic resistance to fungal, oomycete and viral diseases responsible for huge yield losses each year.

Sugarcane

Sugar is extracted from the sweet, juicy stems of sugar cane, and is used worldwide as a sweetener, preservative and in the cosmetics industry.

Species Information

Scientific name: *Saccharum officinarum* L.

Common name: sugar cane, noble cane (English); ikshu, khanda, sarkara (Sanskrit); pundia, paunda (Hindi); poovan karumbu (Tamil)

Conservation status: Not evaluated according to IUCN Red List criteria; widespread in cultivation.

Habitat: Hot humid tropics, in moist soils.

Key Uses: Food and drink, medicine, alcohol production, biofuel, hair removal.

Known hazards: Sugar consumption can be a factor in tooth decay and obesity.

Description

Sugarcane is a tropical, perennial grass that forms lateral shoots at the base to produce multiple stems, typically three to four metres high and about five cm in diametre. The stems grow into cane stalk, which when mature constitutes approximately 75% of the entire plant. A mature stalk is typically composed of 11–16% fibre, 12–16% soluble sugars, 2–3% non-sugars, and 63–73% water. A sugarcane crop is sensitive to the climate, soil type, irrigation, fertilizers, insects, disease control, varieties, and the harvest period. The average yield of cane stalk is 60–70 tonnes per hectare per year. However, this figure can vary between 30 and 180 tonnes per hectare depending on knowledge and crop management approach used in sugarcane cultivation. Sugarcane is a cash crop, but it is also used as livestock fodder.

Cultivation

Sugarcane cultivation requires a tropical or temperate climate, with a minimum of 60 centimetres (24 in) of annual moisture. It is one of the most efficient photosynthesizers in the plant kingdom. It is a C_4 plant, able to convert up to one percent of incident solar energy into biomass. In prime growing regions, such as Mauritius, Dominican Republic, Puerto Rico, India, Indonesia, Pakistan, Peru, Brazil, Bolivia, Colombia, Australia, Ecuador, Cuba, the Philippines, El Salvador and Hawaii, sugarcane crop can produce over 15 kilograms of cane per square metre of sunshine.

Sugarcane is cultivated in the tropics and subtropics in areas with plentiful supply of water, for a continuous period of more than six to seven months each year, either from natural rainfall or through irrigation. The crop does not tolerate severe frosts. Therefore, most of the world's sugarcane is grown between 22°N and 22°S, and some up to 33°N and 33°S. When sugarcane crop is found outside this range, such as the Natal region of South Africa, it is normally due to anomalous climatic conditions in the region, such as warm ocean currents that sweep down the coast. In terms of altitude, sugarcane crop is found up to 1,600 m close to the equator in countries such as Colombia, Ecuador and Peru.

Sugarcane can be grown on many soils ranging from highly fertile well drained mollisols, through heavy cracking vertisols, infertile acid oxisols, peaty histosols to rocky andisols. Both plentiful sunshine and water supplies increase cane production. This has made desert countries with good irrigation facilities such as Egypt as some of the highest yielding sugarcane cultivating regions.

Figure: *Sugarcane flower in Dominica.*

Although sugarcanes produce seeds, modern stem cutting has become the most common reproduction method. Each cutting must contain at least one bud, and the cuttings are sometimes hand-planted. In more technologically advanced countries like the United States and Australia, billet planting is common. Billets harvested from a mechanical harvester are planted by a machine that opens and recloses the ground. Once planted, a stand can be harvested several times; after each harvest, the cane sends up new stalks, called ratoons. Successive harvests give decreasing yields, eventually justifying replanting. Two to 10 harvests are usually made depending on the type of culture. In a country with a mechanical agriculture looking for a high production of large fields, like in North America, sugar canes are replanted after two or three harvests to avoid a lowering in yields. In countries with a more traditional type of agriculture with smaller fields and hand harvesting, like in the French island la Réunion, sugar canes are often harvested up to 10 years before replanting.

Sugarcane is harvested by hand and mechanically. Hand harvesting accounts for more than half of production, and is dominant in the developing world. In hand harvesting, the field is first set on fire. The fire burns dry leaves, and chases or kills any lurking venomous snakes, without harming the stalks and roots. Harvesters then cut the cane just above ground-level using cane knives or machetes. A skilled harvester can cut 500 kilograms (1,100 lb) of sugarcane per hour.

Mechanical harvesting uses a combine, or sugarcane harvester. The Austoft 7000 series, the original modern harvester design, has now been copied by other companies, including Cameco / John Deere. The machine cuts the cane at the base of the stalk, strips the leaves, chops the cane into consistent lengths and deposits it into a transporter following alongside.

The harvester then blows the trash back onto the field. Such machines can harvest 100 long tons (100 t) each hour; however, harvested cane must be rapidly processed. Once cut, sugarcane begins to lose its sugar content, and damage to the cane during mechanical harvesting accelerates this decline. This decline is offset because a modern chopper harvester can complete the harvest faster and more efficiently than hand cutting and loading. Austoft also developed a series of hydraulic high-lift infield transporters to work alongside their harvesters to allow even more rapid transfer of cane to, for example, the nearest railway siding. This mechanical harvesting doesn't require the field to be set on fire; the remains left in the field by the machine consist of the top of the sugar cane and the dead leaves, which act as mulch for the next round of planting.

Pests

The cane beetle (also known as cane grub) can substantially reduce crop yield by eating roots; it can be controlled with imidacloprid (Confidor) or chlorpyrifos (Lorsban). Other important pests are the larvae of some butterfly/moth species, including the turnip moth, the sugarcane borer (*Diatraea saccharalis*), the Mexican rice borer (*Eoreuma loftini*); leaf-cutting ants, termites, spittlebugs (especially *Mahanarva fimbriolata* and *Deois flavopicta*), and the beetle *Migdolus fryanus*. The planthopper insect *Eumetopina flavipes* acts as a virus vector, which causes the sugarcane disease ramu stunt.

Pathogens

Numerous pathogens infect sugarcane, such as sugarcane grassy shoot disease caused by *Phytoplasma*, whiptail disease or sugarcane smut, *pokkah boeng* caused by *Fusarium moniliforme*, Xanthomonas Axonopodis bacteria causes Gumming Disease, and red rot disease caused by *Colletotrichum falcatum*. Viral diseases affecting sugarcane include sugarcane mosaic virus, maize streak virus, and sugarcane yellow leaf virus.

Nitrogen Fixation

Some sugarcane varieties are capable of fixing atmospheric nitrogen in association with the bacterium *Glucoacetobacter diazotrophicus*. Unlike legumes and other nitrogen-fixing plants that form root nodules in the soil in association with bacteria, *G. diazotrophicus* lives within the intercellular spaces of the sugarcane's stem.

Coating seeds with the bacteria is a newly developed technology that can enable every crop species to fix nitrogen for its own use.

Processing

Traditionally, sugarcane processing requires two stages. Mills extract raw sugar from freshly harvested cane, and sometimes bleach it to make "mill white" sugar for local consumption. Refineries, often located nearer to consumers in North America, Europe, and Japan, then produce refined white sugar, which is 99 percent sucrose.

These two stages are slowly merging. Increasing affluence in the sugar-producing tropics increased demand for refined sugar products, driving a trend toward combined milling and refining.

Milling

Sugarcane processing produces cane sugar (sucrose) from sugarcane. Other products of the processing include bagasse, molasses, and filtercake.

Bagasse, the residual dry fibre of the cane after cane juice has been extracted, is used for several purposes:

- fuel for the boilers and kilns,
- production of paper, paperboard products and reconstituted panelboard,

- agricultural mulch, and
- as a raw material for production of chemicals.

The primary use of bagasse and bagasse residue is as a fuel source for the boilers in the generation of process steam in sugar plants. Dried filtercake is used as an animal feed supplement, fertilizer, and source of sugarcane wax.

Molasses is produced in two forms: Blackstrap, which has a characteristic strong flavour because of its vitamin and mineral content, and a purer molasses syrup. Blackstrap molasses is sold as a food and dietary supplement. It is also a common ingredient in animal feed, is used to produce ethanol and rum, and in the manufacturing of citric acid. Purer molasses syrups are sold as molasses, and may also be blended with maple syrup, invert sugars, or corn syrup. Both forms of molasses are used in baking.

Sugar refining further purifies the raw sugar. It is first mixed with heavy syrup and then centrifuged in a process called "affination". Its purpose is to wash away the sugar crystals' outer coating, which is less pure than the crystal interior. The remaining sugar is then dissolved to make a syrup, about 60 percent solids by weight.

The sugar solution is clarified by the addition of phosphoric acid and calcium hydroxide, which combine to precipitate calcium phosphate. The calcium phosphate particles entrap some impurities and absorb others, and then float to the top of the tank, where they can be skimmed off. An alternative to this "phosphatation" technique is "carbonatation", which is similar, but uses carbon dioxide and calcium hydroxide to produce a calcium carbonate precipitate.

After filtering any remaining solids, the clarified syrup is decolorized by filtration through activated carbon. Bone char or coal-based activated carbon is traditionally used in this role. Some remaining colour-forming impurities adsorb to the carbon. The purified syrup is then concentrated to supersaturation and repeatedly crystallized in a vacuum, to produce white refined sugar. As in a sugar mill, the sugar crystals are separated from the molasses by centrifuging. Additional sugar is recovered by blending the remaining syrup with the washings from affination and again crystallizing to produce brown sugar. When no more sugar can be economically recovered, the final molasses still contains 20–30 percent sucrose and 15–25 percent glucose and fructose.

To produce granulated sugar, in which individual grains do not clump, sugar must be dried, first by heating in a rotary dryer, and then by blowing cool air through it for several days.

Ribbon Cane Syrup

Ribbon cane is a subtropical type that was once widely grown in the southern United States, as far north as coastal North Carolina. The juice was extracted with horse or mule-powered crushers; the juice was boiled, like maple syrup, in a flat pan, and then used in the syrup form as a food sweetener. It is not currently a commercial crop, but a few growers find ready sales for their product.

Pollution from Sugarcane Processing

Particulate matter, combustion products, and volatile organic compounds are the primary pollutants emitted during the sugarcane processing. Combustion products include nitrogen oxides (NO_X), carbon monoxide (CO), CO_2, and sulphur oxides (SO_X). Potential emission sources include the sugar granulators, sugar conveying and packaging equipment, bulk loadout operations, boilers, granular carbon and char regeneration kilns, regenerated adsorbent transport systems, kilns and handling equipment (at some facilities), carbonation tanks, multi-effect evaporator stations, and vacuum boiling pans. Modern pollution prevention technologies are capable of addressing all of these potential pollutants.

Production

Brazil led the world in sugarcane production in 2011 with a 734 000 TMT harvest. India was the second largest producer with 342 382 TMT tons, and China the third largest producer with 115 125 TMT tons harvest. The average worldwide yield of sugarcane crops in 2011 was 70.54 tons per hectare. The most productive farms in the world were in Ethiopia with a nationwide average sugarcane crop yield of 126.93 tons per hectare.

The theoretical possible yield for sugar cane, according to 1983 study of Duke, is about 280 metric tons per hectare per year, and small experimental plots in Brazil have demonstrated yields of 236–280 metric tons of fresh cane per hectare. The most promising region for high yield sugarcane production were in sun drenched, irrigated farms of northern Africa, and other deserts with plentiful water from river or irrigation canals.

In the United States, sugarcane is grown commercially in Florida, Hawaii, Louisiana, and Texas.

Brazil uses sugarcane to produce sugar and ethanol for gasoline-ethanol blends (gasohol), a locally popular transportation fuel. In India, sugarcane is used to produce sugar, jaggery and alcoholic beverages.

Cane Ethanol

Ethanol is generally available as a byproduct of sugar production. It can be used as a biofuel alternative to gasoline, and is widely used in cars in Brazil. It is an alternative to gasoline, and may become the primary product of sugarcane processing, rather than sugar.

In Brazil, gasoline is required to contain at least 22 percent bioethanol. This bioethanol is sourced from Brazil's large sugarcane crop.

The production of ethanol from sugar cane is more energy efficient than from corn or sugar beets or palm/vegetable oils, particularly if cane bagasse is used to produce heat and power for the process. Furthermore, if biofuels are used for crop production and transport, the fossil energy input needed for each ethanol energy unit can be very low. EIA estimates that with an integrated sugar cane to ethanol technology, the well-to-wheels CO_2 emissions can be 90 percent lower than conventional gasoline.

Bagasse Applications

Sugarcane is a major crop in many countries. It is one of the plants with the highest bioconversion efficiency. Sugarcane crop is able to efficiently fix solar energy, yielding some 55 tonnes of dry matter per hectare of land annually. After harvest, the crop produces sugar juice and bagasse, the fibrous dry matter. This dry matter is biomass with potential as fuel for energy production.

Sugarcane bagasse is a potentially abundant source of energy for large producers of sugarcane, such as Brazil, India and China. According to one report, with use of latest technologies, bagasse produced annually in Brazil has the potential of meeting 20 percent of Brazil's energy consumption by 2020.

Electricity Production

A number of countries, in particular those devoid of any fossil fuel, have implemented energy conservation and efficiency measures to minimize energy utilized in cane processing and furthermore export any excess electricity to the grid. Bagasse is usually burned to produce steam, which in turn creates electricity. Current technologies, such as those in use in Mauritius, produce over 100 KWh of electricity per tonne of bagasse. With a total world harvest of over 1 billion tonnes of sugar cane per year, the global energy potential from bagasse is over 100,000 GWh. Using Mauritius as a reference, an annual potential of 10,000 GWh of additional electricity could be produced throughout Africa. Electrical generation from bagasse could become quite important, particularly to the rural populations of sugarcane producing nations. Recent cogeneration technology plants are being designed to produce from 200 to over 300 KWh of electricity per tonne of bagasse. As sugarcane is a seasonal crop, shortly after harvest the supply of bagasse would peak, requiring power generation plants to strategically manage the storage of bagasse.

Biogas Production

A greener alternative to burning bagasse for the production of electricity is to convert bagasse into biogas. Technologies are being developed to use enzymes to transform bagasse into advanced biofuel and biogas. Not only could this process realize a greater energy potential, the release of greenhouse gases would be drastically less than simply burning bagasse.

Rubber

Natural rubber, also called India rubber or *caoutchouc*, as initially produced, consists of suitable polymers of the organic compound isoprene, with minor impurities of other organic compounds plus water. Forms of polyisoprene that are useful as natural rubbers are classified as elastomers. Currently, rubber is harvested mainly in the form of the latex from certain trees. The latex is a sticky, milky colloid drawn off by making incisions into the bark and collecting the fluid in vessels in a process called "tapping". The latex then is refined into rubber ready for commercial processing. Natural rubber is used extensively in many applications and products, either alone or in combination with other materials. In most of its useful forms, it has a large stretch ratio, high resilience, and is extremely waterproof.

Varieties

The major commercial source of natural rubber latex is the Pará rubber tree (*Hevea brasiliensis*), a member of the spurge family, *Euphorbiaceae*. This species is widely used because it grows well under cultivation and a properly managed tree responds to wounding by producing more latex for several years.

Many other plants produce forms of latex rich in isoprene polymers, though not all produce usable forms of polymer as easily as the Pará rubber tree does; some of them require more elaborate processing to produce anything like usable rubber, and most are more difficult to tap. Some produce other desirable materials, for example gutta-percha (*Palaquium gutta*) and chicle from *Manilkara* species. Others that have been commercially exploited, or at least have shown promise as sources of rubber, include the rubber fig (*Ficus elastica*), Panama rubber tree (*Castilla elastica*), various spurges (*Euphorbia* spp.), lettuce (*Lactuca* species), the related *Scorzonera tau-saghyz*, various *Taraxacum* species, including common dandelion (*Taraxacum officinale*) and Russian dandelion (*Taraxacum kok-saghyz*), and guayule (*Parthenium argentatum*). The term gum rubber is sometimes applied to the tree-obtained version of natural rubber in order to distinguish it from the synthetic version.

Discovery of Commercial Potential

The Para rubber tree is indigenous to South America. Charles Marie de La Condamine is credited with introducing samples of rubber to the *Académie Royale des Sciences* of France in 1736. In 1751, he presented a paper by François Fresneau to the Académie (eventually published in 1755) which described many of the properties of rubber. This has been referred to as the first scientific paper on rubber. In England, Joseph Priestley, in 1770, observed that a piece of the material was extremely good for rubbing off pencil marks on paper, hence the name "rubber". Later, it slowly made its way around England.

South America remained the main source of the limited amounts of latex rubber used during much of the 19th century. The trade was well protected and exporting seeds from Brazil was said to be a capital offence, although there was no law against it. Nevertheless, in 1876, Henry Wickham smuggled 70,000 Para rubber tree seeds from Brazil and delivered them to Kew Gardens, England. Only 2,400 of these germinated after which the seedlings were then sent to India, Ceylon (Sri Lanka), Indonesia, Singapore, and British Malaya. Malaya (now Peninsular Malaysia) was later to become the biggest producer of rubber. In the early 1900s, the Congo Free State in Africa was also a significant source of natural rubber latex, mostly gathered by forced labour. Liberia and Nigeria also started production of rubber.

In India, commercial cultivation of natural rubber was introduced by the British planters, although the experimental efforts to grow rubber on a commercial scale in India were initiated as early as 1873 at the Botanical Gardens, Calcutta. The first commercial *Hevea* plantations in India were established at Thattekadu in Kerala in 1902.

In Singapore and Malaya, commercial production of rubber was heavily promoted by Sir Henry Nicholas Ridley, who served as the first Scientific Director of the Singapore Botanic Gardens from 1888 to 1911. He distributed rubber seeds to many planters and developed the first technique for tapping trees for latex without causing serious harm to the tree. Because of his very fervent promotion of this crop, he is popularly remembered by the nickname "Mad Ridley".

Properties

Rubber exhibits unique physical and chemical properties. Rubber's stress-strain behaviour exhibits the Mullins effect and the Payne effect, and is often modelled as hyperelastic. Rubber strain crystallizes.

Owing to the presence of a double bond in each repeat unit, natural rubber is susceptible to vulcanisation and sensitive to ozone cracking.

The two main solvents for rubber are turpentine and naphtha (petroleum). The former has been in use since 1764 when François Fresnau made the discovery. Giovanni Fabbroni is credited with the discovery of naphtha as a rubber solvent in 1779. Because rubber does not dissolve easily, the material is finely divided by shredding prior to its immersion.

An ammonia solution can be used to prevent the coagulation of raw latex while it is being transported from its collection site.

Elasticity

On a microscopic scale, relaxed rubber is a disorganized cluster of erratically changing wrinkled chains. In stretched rubber, the chains are almost linear. The restoring force is due to the preponderance of wrinkled conformations over more linear ones.

Cooling below the glass transition temperature still permits local conformational changes but a reordering is practically impossible because of the larger energy barrier for the concerted movement of longer chains. "Frozen" rubber's elasticity is low and strain results from small changes of bond lengths and angles. (This caused the Challenger disaster, where flattened o-rings failed to relax to fill a widening gap.) The glass transition is fast and reversible: the force resumes on heating.

The parallel chains of stretched rubber are susceptible to crystallization. This takes some time because turns of twisted chains have to move out of the way of the growing crystallites. Crystallization has occurred, for example, when, after days, an inflated toy balloon is found withered at a relatively large remaining volume. Where it is touched, it shrinks because the temperature of the hand is enough to melt the crystals.

Vulcanization of rubber creates disulfide bonds between chains, which limits the degrees of freedom and results in chains that tighten more quickly for a given strain, thereby increasing the elastic force constant and making the rubber harder and less extensible.

Chemical Makeup

Latex is the polymer cis-1,4-polyisoprene – with a molecular weight of 100,000 to 1,000,000 daltons. Typically, a small percentage (up to 5% of dry mass) of other materials, such as proteins, fatty acids, resins, and inorganic materials (salts) are found in natural rubber. Polyisoprene can also be created synthetically, producing what is sometimes referred to as "synthetic natural rubber", but the synthetic and natural routes are completely different.

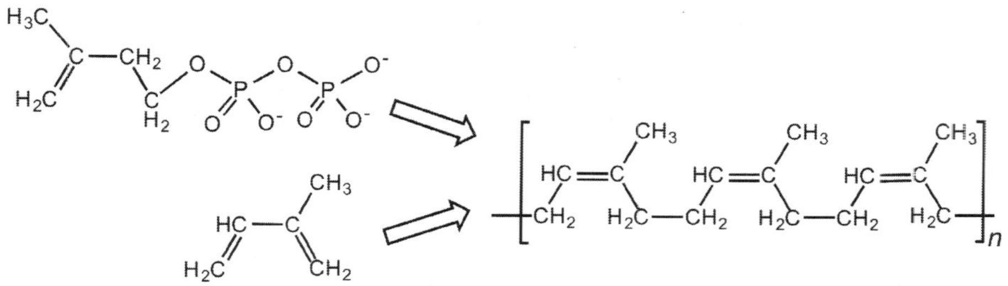

Figure: *Chemical structure of cis-polyisoprene, the main constituent of natural rubber. Synthetic cis-polyisoprene and natural cis-polyisoprene are derived from different precursors, isopentenyl pyrophosphate and isoprene.*

Some natural rubber sources, such as gutta-percha, are composed of trans-1,4-polyisoprene, a structural isomer that has similar, but not identical, properties.

Natural rubber is an elastomer and a thermoplastic. Once the rubber is vulcanized, it will turn into a thermoset. Most rubber in everyday use is vulcanized to a point where it shares properties of both; i.e., if it is heated and cooled, it is degraded but not destroyed.

The final properties of a rubber item depend not just on the polymer, but also on modifiers and fillers, such as carbon black, factice, whiting, and a host of others.

Biosynthesis

Rubber particles are formed in the cytoplasm of specialized latex-producing cells called laticifers within rubber plants. Rubber particles are surrounded by a single phospholipid membrane with hydrophobic tails pointed inward. The membrane allows biosynthetic proteins to be sequestered at the surface of the growing rubber particle, which allows new monomeric units to be added from outside the biomembrane, but within the lacticifer. The rubber particle is an enzymatically active entity that contains three layers of material, the rubber particle, a biomembrane, and free monomeric units. The biomembrane is held tightly to the rubber core due to the high negative charge along the double bonds of the rubber polymer backbone. Free monomeric units and conjugated proteins make up the outer layer. The rubber precursor is isopentenyl pyrophosphate (an allylic compound), which elongates by Mg^{2+}-dependent condensation by the action of rubber transferase. The monomer adds to the pyrophosphate end of the growing polymer. The process displaces the terminal high-energy pyrophosphate. The reaction produces a cis polymer.

The initiation step is catalyzed by prenyltransferase, which converts three monomers of isopentenyl pyrophosphate into farnesyl pyrophosphate. The farnesyl pyrophosphate can bind to rubber transferase to elongate a new rubber polymer.

The required isopentenyl pyrophosphate is obtained from the mevalonate pathway, which is derives from acetyl-CoA in the cytosol. In plants, isoprene pyrophosphate can also be obtained from 1-deox-D-xyulose-5-phosphate/2-C-methyl-D-erythritol-4-phosphate pathway within plasmids. The relative ratio of the farnesyl pyrophosphate initiator unit and isoprenyl pyrophosphate elongation monomer determines the rate of new particle synthesis versus elongation of existing particles. Though rubber is known to be produced by only one enzyme, extracts of latex have shown numerous small molecular weight proteins with unknown function. The proteins possibly serve as cofactors, as the synthetic rate decreases with complete removal.

Current Sources

Close to 21 million tons of rubber were produced in 2005, of which approximately 42% was natural. Since the bulk of the rubber produced is of the synthetic variety, which is derived from petroleum, the price of natural rubber is determined, to a large extent, by the prevailing global price of crude oil. Today, Asia is the main source of natural rubber, accounting for about 94% of output in 2005. The three largest producing countries, Thailand, Indonesia (2.4m tons) and Malaysia, together account for around 72% of all natural rubber production. Natural rubber is not cultivated widely in its native continent of South America due to the existence of South American leaf blight, and other natural predators of the rubber tree.

Cultivation

Rubber latex is extracted from rubber trees. The economic life period of rubber trees in plantations is around 32 years – up to 7 years of immature phase and about 25 years of productive phase.

The soil requirement of the plant is generally well-drained, weathered soil consisting of laterite, lateritic types, sedimentary types, nonlateritic red, or alluvial soils.

The climatic conditions for optimum growth of rubber trees are:

- Rainfall of around 250 cm evenly distributed without any marked dry season and with at least 100 rainy days per year
- Temperature range of about 20 to 34°C, with a monthly mean of 25 to 28°C
- High atmospheric humidity of around 80%
- Bright sunshine amounting to about 2000 hours per year at the rate of six hours per day throughout the year
- Absence of strong winds

Many high-yielding clones have been developed for commercial planting. These clones yield more than 2,000 kg of dry rubber per hectare per year, when grown under ideal conditions.

Collection

In places such as Kerala, where coconuts are in abundance, the half shell of coconut is used as the collection container for the latex, but glazed pottery or aluminium or plastic cups are more common elsewhere. The cups are supported by a wire that encircles the tree. This wire incorporates a spring so it can stretch as the tree grows. The latex is led into the cup by a galvanised "spout" knocked into the bark. Tapping normally takes place early in the morning, when the internal pressure of the tree is highest. A good tapper can tap a tree every 20 seconds on a standard half-spiral system, and a common daily "task" size is between 450 and 650 trees. Trees are usually tapped on alternate or third days, although many variations in timing, length, and number of cuts are used.

The latex, which contains 25–40% dry rubber, is in the bark, so the tapper must avoid cutting right through to the wood, else the growing cambial layer will be damaged and the renewing bark will be badly deformed, making later tapping difficult. It is usual to tap a pannel at least twice, sometimes three times, during the tree's life. The economic life of the tree depends on how well the tapping is carried out, as the critical factor is bark consumption. A standard in Malaysia for alternate daily tapping is 25 cm (vertical) bark consumption per year. The latex tubes in the bark ascend in a spiral to the right. For this reason, tapping cuts usually ascend to the left to cut more tubes.

The trees will drip latex for about four hours, stopping as latex coagulates naturally on the tapping cut, thus blocking the latex tubes in the bark. Tappers usually rest and have a meal after finishing their tapping work, then start collecting the liquid "field latex" at about midday. Some trees will continue to drip after the collection and this leads to a small amount of "cup lump" which is collected at the next tapping. The latex that coagulates on the cut is also collected as "tree lace". Tree lace and cup lump together account for 10–20% of the dry rubber produced. Latex that drips onto the ground, "earth scrap", is also collected periodically for processing of low-grade product.

Field Coagula

There are four types of field coagula, "cuplump", "treelace", "smallholders' lump" and "earth scrap". Each has significantly different properties.

Cuplump is the coagulated material found in the collection cup when the tapper next visits the tree to tap it again. It arises from latex clinging to the walls of the cup after the latex was last poured into the bucket, and from late-dripping latex exuded before the latex-carrying vessels of the tree become blocked. It is of higher purity and of greater value than the other three types.

Treelace is the coagulum strip that the tapper peels off the previous cut before making a new cut. It usually has higher copper and manganese contents than cuplump. Both copper and manganese are pro-oxidants and can lower the physical properties of the dry rubber.

Smallholders' lump is produced by smallholders who collect rubber from trees a long way away from the nearest factory. Many Indonesian smallholders, who grow paddy in remote areas, tap dispersed trees on their way to work in the paddy fields and collect the latex (or the coagulated latex) on their way home. As it is often impossible to preserve the latex sufficiently to get it to a factory that processes latex in time for it to be used to make high quality products, and as the latex would anyway have coagulated by the time it reached the factory, the smallholder will coagulate it by any means available, in any container available. Some smallholders use small containers, buckets etc., but often the latex is coagulated in holes in the ground, which are usually (but not always) lined with plastic. Acidic materials and fermented fruit juices are used to coagulate the latex – a form of assisted biological coagulation. Little care is taken to exclude twigs, leaves, and even bark from the lumps that are formed, which may also include treelace collected by the smallholder.

Earth scrap is the material that gathers around the base of the tree. It arises from latex overflowing from the cut and running down the bark of the tree, from rain flooding a collection cup containing latex, and from spillage from tappers' buckets during collection. It contains soil and other contaminants, and has variable rubber content depending on the amount of contaminants mixed with it. Earth scrap is collected by the field workers two or three times a year and may be cleaned in a scrap-washer to recover the rubber, or sold off to a contractor who will clean it and recover the rubber. It is of very low quality and under no circumstances should it be included in block rubber or brown crepe.

Processing

The latex will coagulate in the cups if kept for long. The latex has to be collected before coagulation. The collected latex, "field latex", is transferred into coagulation tanks for the preparation of dry rubber or transferred into air-tight containers with sieving for ammoniation. Ammoniation is necessary to preserve the latex in a colloidal state for longer periods of time.

Latex is generally processed into either latex concentrate for manufacture of dipped goods or it can be coagulated under controlled, clean conditions using formic acid. The coagulated latex can then be processed into the higher-grade, technically specified block rubbers such as SVR 3L or SVR CV or used to produce Ribbed Smoke Sheet grades.

Naturally coagulated rubber (cup lump) is used in the manufacture of TSR10 and TSR20 grade rubbers. The processing of the rubber for these grades is a size reduction and cleaning process to remove contamination and prepare the material for the final stage of drying.

The dried material is then baled and palletized for storage and shipment in various methods of transportation.

Transportation

Natural rubber latex is shipped from factories in south-west Asia, South America, and North Africa to destinations around the world. As the cost of natural rubber has risen significantly, the shipping methods which offer the lowest cost per unit of weight are preferred. Depending on the destination, warehouse availability, and transportation conditions, some methods are more suitable to certain buyers than others. In international trade, latex rubber is mostly shipped in 20-foot ocean containers. Inside the ocean container, various types of smaller containers are used by factories to store latex rubber.

Uses

Contemporary Manufacturing: Around 25 million tonnes of rubber is produced each year, of which 42 percent is natural rubber. The remainder is synthetic rubber derived from petrochemical sources. Around 70 percent of the world's natural rubber is used in tires. The top end of latex production results in latex products such as surgeons' gloves, condoms, balloons and other relatively high-value products. The mid-range which comes from the technically-specified natural rubber materials ends up largely in tires but also in conveyor belts, marine products, windshield wipers and miscellaneous rubber goods.

Natural rubber offers good elasticity, while synthetic materials tend to offer better resistance to environmental factors such as oils, temperature, chemicals or ultraviolet light and suchlike. "Cured rubber" is rubber which has been compounded and subjected to the vulcanisation process which creates cross-links within the rubber matrix.

Natural rubber is susceptible to degradation by a wide range of bacteria.

Oil Palm

Elaeis guineensis is a species of palm commonly called African oil palm or macaw-fat. It is the principal source of palm oil. It is native to west and southwest Africa, specifically the area between Angola and the Gambia; the species name *guineensis* refers to the name for the area, Guinea, and not the modern country which now bears that name.

The species is also now naturalised in Madagascar, Sri Lanka, Malaysia, Sumatra, Central America, the West Indies and several islands in the Indian and Pacific Oceans. The closely related American oil palm *Elaeis oleifera* and a more distantly related palm, *Attalea maripa*, are also used to produce palm oil.

Human use of oil palms may date as far back as 5,000 years in West Africa; in the late 1800s, archaeologists discovered palm oil in a tomb at Abydos dating back to 3,000 BCE. It is thought that Arab traders brought the oil palm to Egypt.

Description

Mature palms are single-stemmed and grow to 20 m tall. The leaves are pinnate and reach between 3-5 m long. A young palm produces about 30 leaves a year. Established palms over 10 years produce about 20 leaves a year. The flowers are produced in dense clusters; each individual flower is small, with three sepals and three petals.

The palm fruit takes five to six months to mature from pollination to maturity. It is reddish, about the size of a large plum, and grows in large bunches.

Each fruit is made up of an oily, fleshy outer layer (the pericarp), with a single seed (the palm kernel), also rich in oil. When ripe, each bunch of fruit weighs 40–50 kg (88–110 lb).

Planting

For each hectare of oil palm, which is harvested year-round, the annual production averages 10 tonnes of fruit yielding 4,000 kg of palm oil and 750 kg of seed kernels yielding 500 kg of high quality palm kernel oil, as well as 600 kg of kernel meal. Palm fronds and kernel meal are processed for use as livestock feed.

All modern, commercial planting material consists of tenera palms or DxP hybrids, which are obtained by crossing thickshelled dura with shell-less pisifera. Although common commercial germinated seed is as thick-shelled as the dura mother palm, the resulting palm will produce thin-shelled tenera fruit. An alternative to germinated seed, once constraints to mass production are overcome, are tissue-cultured or "clonal" palms, which provide "true copies" of high-yielding DxP palms.

An oil palm nursery must have an uninterrupted supply of clean water and topsoil which is both well-structured and sufficiently deep to accommodate three rounds of on-site bag-filling. Approximately 35 ha can grow enough seedlings over a three-year period to plant a 5,000-ha plantation. Prenursery seedlings must be watered daily. Whenever rainfall is less than 10 mm per day, irrigation is required, and the system must be capable of uniformly applying 6.5 mm water per day.

Prenursery seedlings in the four-leaf stage of development (10 to 14 weeks after planting) are usually transplanted to the main nursery after their gradual adjustment to full sunlight and a rigid selection process. During culling, seedlings that have grassy, crinkled, twisted, or rolled leaves are discarded.

Weeds growing in the polybags must be carefully pulled out. Herbicides should not be used. Numerous insects (ants, armyworms, bagworms, aphids, thrips, mites, grasshoppers, and mealybugs) and vertebrates (rats, squirrels, porcupines, wild boar, and monkeys) are pests in oil palm nurseries and must be carefully identified before control measures are implemented.

After eight months in the nursery, normal healthy plants should be 0.8–1 m in height and display five to eight functional leaves.

The proper approach to oil palm development begins with the establishment of leguminous cover plants, immediately following land clearing. They help prevent soil erosion and surface run-off, improve soil structure and palm root development, increase the response to mineral fertilizer in later years, and reduce the danger of micronutrient deficiencies. Leguminous cover plants also help prevent outbreaks of *Oryctes* beetles, which nest in exposed decomposing vegetation. Both phosphorus and potassium fertilizers are needed to maximize the leguminous cover plants' symbiotic nitrogen-fixation potential of approximately 200 kg nitrogen/ha/yr, and are applied to most soils at 115 to 300 kg phosphorus oxide/ha and 35 to 60 kg potassium oxide/ha. Young palms are severely set back where grasses are allowed to dominate the inter-row vegetation, particularly on poor soils where the correction of nutrient deficiencies is difficult and costly.

Crop Nutrient

Figure: *Oil palm* Elaeis guinnensis

Nutrient uptake is low during the first year but increases steeply between year one and year three (when harvesting commences) and stabilizes around years five to six. Early applications of fertilizer, better planting material, and more rigid culling have led to a dramatic increase in early yields in the third to sixth years from time of planting. In regions without a significant drop in rainfall, yields of over 25 tonnes of fresh fruit bunches per hectare have been achieved in the second year of harvesting.

Nitrogen deficiency is usually associated with conditions of water-logging, heavy weed infestation, and topsoil erosion. Symptoms are a general paling and stiffening of the pinnae, which lose their glossy lustre. Extended deficiency will reduce the number of effective fruit bunches produced, as well as the bunch size.

Phosphorus-deficient leaves do not show specific symptoms, but frond length, bunch size, and trunk diametre are all reduced.

Potassium deficiency is very common and is the major yield constraint in sandy or peaty soils. The most frequent symptom is "confluent orange spotting". Pale green spots appear on the pinnae of older leaves; as the deficiency intensifies, the spots turn orange or reddish-orange and desiccation sets in, starting from the tips and outer margins of the pinnae. Other symptoms are "orange blotch" and "midcrown yellowing". In soils having a low water-holding capacity (sands and peats), potassium deficiency can lead to a rapid, premature desiccation of fronds.

Copper deficiency is common on deep peat soils and occurs also on very sandy soils. It appears initially as whitish-yellow mottling of younger fronds. As the deficiency intensifies, yellow, mottled, interveinal stripes appear, and rusty, brown spots develop on the distal ends of leaflets. Affected fronds and leaflets are stunted and leaflets dry up. On sandy soils, palms recover rapidly after a basal application of 50 grams of copper sulphate. On peat soils, lasting correction of copper deficiency is difficult, as applied copper sulphate is rendered unavailable. A promising method of correcting copper deficiency on peat soil is to mix copper sulphate with clay soil and to form tennis-ball sized "copper mudballs" that are placed around the palm to provide a slow-release source of available copper.

Healthy, well selected seedlings are necessary for early and sustained high yield. In most cases, granular multinutrient compound fertilizers are the preferred nutrient source for seedlings in the nursery. Where subsoil is used to fill the polybags, extra dressings of Kieserite may be required (10-15 g every six to eight weeks). Where compound fertilizers are not available, equivalent quantities of straight materials should be used.

To maintain good fertilizer response and high yields in older palms, selective thinning is often necessary.

Cross-breeding

Figure: *Oil palm fruit*

Unlike other relatives, oil palms do not produce offshoots; propagation is by sowing the seeds.

Several varieties and forms of *Elaeis guineensis* have been selected that have different characteristics. These include:

- *E. guineensis* fo. *dura*
- *E. guineensis* var. *pisifera*
- *E, guineensis* fo. *tenera*

Before the Second World War, selection work had started in the *Deli dura* population in Malaya. Pollen was imported from Africa, and DxT and DxP crosses were made. Segregation of fruit forms in crosses made in the 1950s was often incorrect. In the absence of a good marker gene, there was no way of knowing whether control of pollination was adequate.

After the work of Beirnaert and Vanderweyen (1941), it became feasible to monitor the efficacy of controlled pollination. From 1963 until the introduction of the palm-pollinating weevil *Elaeidobius kamerunicus* in 1982, contamination in Malaysia's commercial plantings was generally low. Thrips, the main pollinating agent at that time, apparently rarely gained access to bagged female inflorescences. However, *E. kamerunicus* is much more persistent, and after it was introduced, *Deli dura* contamination became a significant problem. This problem apparently persisted for much of the 1980s, but in a 1991 comparison of seed sources, contamination had been reduced to below 2%, indicating control had been restored.

A 1992 study at a trial plot in Banting, Selangor, revealed the "yield of Deli dura oil palms after four generations of selection was 60% greater than that of the unselected base population. Crossing the dura and pisifera to give the thin-shelled tenera fruit type improved partitioning of dry matter within the fruit, giving a 30% increase in oil yield at the expense of shell, without changing total dry matter production."

Disease

Basal stem rot (BSR), caused by the fungus *Ganoderma*, is the most serious disease of oil palm in Malaysia and Indonesia. Previously, research on basal stem rot was hampered by the failure to artificially infect oil palms with the fungus. Although *Ganoderma* had been associated with BSR, proof of its pathogenicity to satisfy Koch's postulate was only achieved in the early 1990s by inoculating oil palm seedling roots or by using rubber wood blocks. A reliable and quick technique was developed for testing the pathogenicity of the fungus by inoculating oil palm germinated seeds.

This fatal disease can lead to losses as much as 80% after repeated planting cycles. *Ganoderma* produces enzymes that degrade the infected xylem, thus causing serious problems to the distribution of water and other nutrients to the top of the palm. *Ganoderma* infection is well defined by its lesion in the stem. The cross-section of infected palm stem shows that the lesion appears as a light brown area of rotting tissue with a distinctive, irregularly shaped, darker band at the borders of this area. The infected tissue become as an ashen-grey powdery and if the palm remains standing, the infected trunk rapidly become hollow.

In a 2007 study in Portugal, scientists suggested control of the fungus on oil palms would benefit from further consideration of the process as one of white rot. *Ganoderma* is an extraordinary organism capable exclusively of degrading lignin to carbon dioxide and water; celluloses are then available as nutrients for the fungus. It is necessary to consider this mode of attack as a white rot involving lignin biodegradation, for integrated control. The existing literature does not report this area and appears to be concerned particularly with the mode of spread and molecular biology of *Ganoderma*. The white rot perception opens up new fields in breeding/selecting for resistant cultivars of oil palms with high lignin content, ensuring the conditions for lignin decomposition are reduced, and simply sealing damaged oil palms to stop decay. The spread likely is by spores rather than roots. The knowledge gained can be employed in the rapid degradation of oil palm waste on the plantation floor by inoculating suitable fungi, and/or treating the waste more appropriately (e.g. chipping and spreading over the floor rather than windrowing).

Endophytic bacteria are organisms inhabiting plant organs that at some time in their life cycles can colonize the internal plant tissues without causing apparent harm to the host. Introducing endophytic bacteria to the roots to control plant disease is to manipulate the indigenous bacterial communities of the roots in a manner, which leads to enhanced suppression of soil-borne pathogens. The use of endophytic bacteria should thus be preferred to other biological control agents, as they are internal colonizers, with better ability to compete within the vascular systems, limiting *Ganoderma* for both nutrients and space during its proliferation. Two bacterial isolates, *Burkholderia cepacia*(B3) and *Pseudomonas aeruginosa*(P3) were selected for evaluation in the glasshouse for their efficacy in enhancing growth and subsequent suppression of the spread of BSR in oil palm seedlings.

Little leaf syndrome has not been fully explained, but has often been confused with boron deficiency. The growing point is damaged, sometimes by *Oryctes* beetles. Small, distorted leaves resembling a boron deficiency emerge. This is often followed by secondary pathogenic infections in the spear that can lead to spear rot and palm death.

History

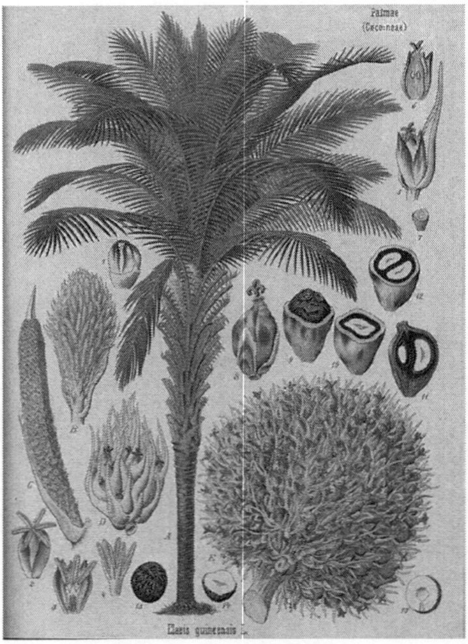

Figure: *African oil palm (*Elaeis guineensis*)*

Elaeis guineensis originated in Guinea, Africa and was first illustrated by Nicholaas Jacquin in 1763.

Oil palms were introduced to Java by the Dutch in 1848, and to Malaysia (then the British colony of Malaya) in 1910 by Scotsman William Sime and English banker Henry Darby. The species of palm tree *Elaeis guineensis* was taken to Malaysia from Eastern Nigeria in 1961. As noted it originally grew in West Africa. The southern coast of Nigeria was originally called the Palm oil coast by the first Europeans who arrived there and traded in the commodity. This area was later renamed the Bight of Biafra.

Malaysia

In Malaysia, the first plantations were mostly established and operated by British plantation owners, such as Sime Darby and Boustead, and remained listed in London until the Malaysian government engineered their "Malaysianisation" throughout the 1960s and 1970s.

Federal Land Development Authority (Felda) is the world's biggest oil palm planter, with planted area close to 900,000 hectares in Malaysia and Indonesia. Felda was formed on July 1, 1956 when the Land Development Act came into force with the main aim of eradicating poverty. Settlers were each allocated 10 acres of land (about 4 hectares) planted either with oil palm or rubber, and given 20 years to pay off the debt for the land.

After Malaysia achieved independence in 1957, the government focused on value-added of rubber planting, boosting exports, and alleviating poverty through land schemes. In the 1960s and 1970s, the government encouraged planting of other crops, to cushion the economy when world prices of tin and rubber plunged. Rubber estates gave way to oil palm plantations. In 1961, Felda's first oil palm settlement opened, with 3.75 km^2 of land. As of 2000, 6855.2 km^2 (approximately 76%) of the land under Felda's programmes were devoted to oil palms. By 2008, Felda's resettlement broadened to 112,635 families, who work on 8533.13 km^2 of agriculture land throughout Malaysia. Oil palm planting took up 84% of Felda's plantation landbank.

FELDA's success led to the establishment of other development schemes to support the establishment of small-farmer oil palm cultivation. The Federal Land Consolidation and Rehabilitation Authority (FELCRA) was established in 1966 and the Sarawak Land Consolidation and Rehabilitation Authority (SALCRA) was formed in 1976. The primary objective of these organizations is to assist in the development of rural communities and reduce poverty through the cultivation of high yielding crops such as palm oil.

As of November 2011, SALCRA had developed 18 estates totalling approximately 51,000 hectares. That year the organization shared dividends with 16,374 landowners participating in the programme.

Palm Oil Production

Oil is extracted from both the pulp of the fruit (palm oil, an edible oil) and the kernel (palm kernel oil, used in foods and for soap manufacture). For every 100 kg of fruit bunches, typically 22 kg of palm oil and 1.6 kg of palm kernel oil can be extracted.

The high oil yield of oil palms (as high as 7,250 litres per hectare per year) has made it a common cooking ingredient in Southeast Asia and the tropical belt of Africa. Its increasing use in the commercial food industry in other parts of the world is buoyed by its cheaper pricing, the high oxidative stability of the refined product, and high levels of natural antioxidants.

The oil palm originated in West Africa, but has since been planted successfully in tropical regions within 20 degrees of the equator. In the Republic of the Congo, or Congo Brazzaville, precisely in the Northern part, not far from Ouesso, local people produce this oil by hand. They harvest the fruit, boil it to let the water evaporate, then press what is left to collect the reddish-orange-coloured oil.

In 1995, Malaysia was the world's largest producer, with a 51% of world share, but since 2007, Indonesia has been the world's largest producer, supplying approximately 50% of world palm oil volume.

Worldwide palm oil production for season 2011/2012 was 50.3 million metric tons, increasing to 52.3 million tons for 2012/13. In 2010/2011, total production of palm kernels was 12.6 million tonnes.

The Urhobo people of Nigeria use the extract to make Amiedi soup.

Oil Palm Research

Key scientific journals publishing on oil palms and related topics include:

- *Journal of Applied Polymer Science*
- *Conservation Letters*
- *Bioresource Technology*
- *Trends in Ecology and Evolution*

Social and Environmental Impacts

The social and environmental impacts of oil palm cultivation is a highly controversial topic. Oil palm is a valuable economic crop and provides a major source of employment. It allows many small landholders to participate in the cash economy and often results in the upgrade of the infrastructure (schools, roads, telecommunications) within that area. However, there are cases where native customary lands have been appropriated by oil palm plantations without any form of consultation or compensation, leading to social conflict between the plantations and local residents. In some cases, oil palm plantations are dependent on imported labour or illegal immigrants, with some concerns about the employment conditions and social impacts of these practices.

Biodiversity loss (including the potential extinction of charismatic species) is one of the most serious negative effects of oil palm cultivation. Large areas of already threatened tropical rainforest are often cleared to make way for palm oil plantations, especially in Southeast Asia, where enforcement of forest protection laws is lacking. In some states where oil palm is established, lax enforcement of environmental legislation leads to encroachment of plantations into protected areas, encroachment into riparian strips, open burning of plantation wastes, and release of palm mill pollutants such as palm oil mill effluent (POME) in the environment. Some of these states have recognised the need for increased environmental protection, resulting in more environment-friendly practices. Among those approaches is anaerobic treatment of POME, which can be a good source for biogas (methane) production and electricity generation. Anaerobic treatment of POME has been practiced in Malaysia and Indonesia. Like most wastewater sludge, anaerobic treatment of POME results in dominance of *Methanosaeta concilii*. It plays an important role in methane production from acetate, and the optimum condition for its growth should be considered to harvest biogas as renewable fuel.

Demand for palm oil has increased in recent years due to its use as a biofuel, but recognition that this increases the environmental impact of cultivation, as well as causing a food vs fuel issue, has forced some developed nations to reconsider their policies on biofuel to improve standards and ensure sustainability. However, critics point out that even companies signed up to the Roundtable on Sustainable Palm Oil continue to engage in environmentally damaging practices and that using palm oil as biofuel is perverse because it encourages the conversion of natural habitats such as forests and peatlands, releasing large quantities of greenhouse gases.

Carbon Balance

Oil palm production has been documented as a cause of substantial and often irreversible damage to the natural environment. Its impacts include deforestation, habitat loss of critically endangered species, and a significant increase in greenhouse gas emissions.

The pollution is exacerbated because many rainforests in Indonesia and Malaysia lie atop peat bogs that store great quantities of carbon, which are released when the forests are cut down and the bogs are drained to make way for the plantations.

Environmental groups, such as Greenpeace, claim the deforestation caused by making way for oil palm plantations is far more damaging for the climate than the benefits gained by switching to biofuel. Fresh land clearances, especially in Borneo, are contentious for their environmental impact. Despite thousands of square kilometres of land standing unplanted in Indonesia, tropical hardwood forests are being cleared for palm oil plantations. Furthermore, as the remaining unprotected lowland forest dwindles, developers are looking to plant peat swamp land, using drainage that begins an oxidation process of the peat which can release 5,000 to 10,000 years worth of stored carbon. Drained peat is also at very high risk of forest fire. There is a clear record of fire being used to clear vegetation for oil palm development in Indonesia, where in recent years drought and man-made clearances have led to massive uncontrolled forest fires, covering parts of Southeast Asia in haze and leading to an international crisis with Malaysia. These fires have been blamed on a government with little ability to enforce its own laws, while impoverished small farmers and large plantation owners illegally burn and clear forests and peat lands to develop the land rather than reap the environmental benefits it could offer.

Many of the major companies in the vegetable oil economy participate in the Roundtable on Sustainable Palm Oil, which is trying to address this problem. For example, in 2008, Unilever, a member of the group, committed to use only oil palm oil which is certified as sustainable, by ensuring the large companies and smallholders that supply it convert to sustainable production by 2015.

Meanwhile, much of the recent investment in new palm plantations for biofuel has been funded through carbon credit projects through the Clean Development Mechanism; however, the reputational risk associated with the unsustainable palm plantations in Indonesia has now made many funds wary of such investment.

Palm Biomass as Fuel

Some scientists and companies are going beyond using just the oil, and are proposing to convert fronds, empty fruit bunches and palm kernel shells harvested from oil palm plantations into renewable electricity, cellulosic ethanol, biogas, biohydrogen and bioplastic. Thus, by using both the biomass from the plantation as well as the processing residues from palm oil production (fibres, kernel shells, palm oil mill effluent), bioenergy from palm plantations can have an effect on reducing greenhouse gas emissions. Examples of these production techniques have been registered as projects under the Kyoto Protocol's Clean Development Mechanism.

By using palm biomass to generate renewable energy, fuels and biodegradable products, both the energy balance and the greenhouse gas emissions balance for palm biodiesel is improved. For every tonne of palm oil produced from fresh fruit bunches, a farmer harvests around 6 tonnes of waste palm fronds, 1 tonne of palm trunks, 5 tonnes of empty fruit bunches, 1 tonne of press fibre (from the mesocarp of the fruit), half a tonne of palm kernel endocarp, 250 kg of palm kernel press cake, and 100 tonnes of palm oil mill effluent. Some oil palm plantations incinerate biomass to generate power for palm oil mills. Some other oil palm plantations yield large amount of biomass that can be recycled into medium density fibreboards and light furniture. In efforts to reduce greenhouse gas emissions, scientists treat palm oil mill effluent to extract biogas. After purification, biogas can substitute for natural gas for use at factories. Anaerobic treatment of palm oil mill effluent, practiced in Malaysia and Indonesia, results in domination of *Methanosaeta concilii*. It plays an important role in methane production from acetate and the optimum condition for its growth should be considered to harvest biogas as renewable fuel.

Unfortunately, the production of palm oil has detrimental effects on the environment and is not considered to be a sustainable biofuel. The deforestation occurring throughout Malaysia and Indonesia as a result of the growing demand for this plant has made scarce natural habitats for orangutans and other rainforest dwellers. More carbon is released during the life cycle of a palm oil plant to its use as a biofuel than is emitted by the same volume of fossil fuels.

Malayan Folkculture

Since the days when the 'guineesis' was first introduced by the British, Indian labourers were brought in to work the estates. There, Hindu beliefs mixed with the local Malay culture and started the usage of palm seeds by traditional healers suffixed with *tok 'bomoh'* or *pawang* in the local language. Every bunch of palm fruit usually bears a single 'illustrious' seed which looks like a shiny black pearl called *sbatmi* in Tamil and *shakti* in Malay. These are used as accessories by the *bomoh* and *pawang* in the mixed ritual for peace with nature as these are believed to contain mystical healing properties, and those wearing them are blessed by nature.

Modern usage has seen more common people keeping these as charm or fashion items to feel at peace, owing to their use by celebrities. All palm seeds contain acid; these *sbatmi* are no different and should be handled with care. *Sbatmi* lost some popularity when it was used in a grisly ritual by Mona Fandey in 1993.

Cashew

The cashew tree is a tropical evergreen that produces the cashew nut and the cashew apple. Officially classed as *Anacardium occidentale*, it can grow as high as 14 metres (46 ft), but the dwarf cashew, growing up to 6 metres (20 ft), has proved more profitable, with earlier maturity and higher yields.

The cashew nut is served as a snack or used in recipes, like other nuts, although it is actually a seed. The cashew apple is a fruit, whose pulp can be processed into a sweet, astringent fruit drink or distilled into liqueur.

The shell of the cashew nut yield derivatives that can be used in many applications from lubricants to paints, and other parts of the tree have traditionally been used for snake-bites and other folk remedies.

Originally native to northeastern Brazil, the tree is now widely grown in tropical regions, Vietnam and Nigeria being major producers, in addition to India, the Ivory Coast, Indonesia.

Habitat and Growth

Figure: *'Anacardium occidentale', from Koehler's 'Medicinal-Plants' (1887)*

The tree is large and evergreen, growing to 10-12m (~32 ft) tall, with a short, often irregularly shaped trunk. The leaves are spirally arranged, leathery textured, elliptic to obovate, 4 to 22 cm long and 2 to 15 cm broad, with smooth margins. The flowers are produced in a panicle or corymb up to 26 cm long; each flower is small, pale green at first, then turning reddish, with five slender, acute petals 7 to 15 mm long. The largest cashew tree in the world covers an area of about 7,500 square metres (81,000 sq ft); it's located in Natal, Rio Grande do Norte, Brazil, and Philippines

The fruit of the cashew tree is an accessory fruit (sometimes called a pseudocarp or false fruit). What appears to be the fruit is an oval or pear-shaped structure, a hypocarpium, that develops from the pedicel and the receptacle of the cashew flower. Called the cashew apple, better known in Central America as *marañón*, it ripens into a yellow and/or red structure about 5–11 cm long. It is edible, and has a strong "sweet" smell and a sweet taste.

The pulp of the cashew apple is very juicy, but the skin is fragile, making it unsuitable for transport. In Latin America, a fruit drink is made from the cashew apple pulp which has a very refreshing taste and tropical flavour that can be described as having notes of mango, raw green pepper, and just a little hint of grapefruit-like citrus.

The true fruit of the cashew tree is a kidney or boxing-glove shaped drupe that grows at the end of the cashew apple. The drupe develops first on the tree, and then the pedicel expands to become the cashew apple. Within the true fruit is a single seed, the cashew nut. Although a nut in the culinary sense, in the botanical sense the nut of the cashew is a seed. The seed is surrounded by a double shell containing an allergenic phenolic resin, anacardic acid, a potent skin irritant chemically related to the better-known allergenic oil urushiol which is also a toxin found in the related poison ivy.

Properly roasting cashews destroys the toxin, but it must be done outdoors as the smoke (not unlike that from burning poison ivy) contains urushiol droplets which can cause severe, sometimes life-threatening, reactions by irritating the lungs. People who are allergic to cashew urushiols may also react to mango or pistachio which are also in the Anacardiaceae family. Some people are allergic to cashew nuts, but cashews are a less frequent allergen than other nuts or peanuts.

Dispersal

While the cashew plant is native to northeast Brazil, the Portuguese took it to Goa, India, between 1560 and 1565. From there it spread throughout Southeast Asia and eventually Africa.

Cashew Nut

Cashew nuts are a popular snack and food source. Cashews, unlike other oily tree nuts, contain starch to about 10% of their weight. This makes them more effective than other nuts in thickening water-based dishes such as soups, meat stews, and some Indian milk-based desserts. Many southeast Asian and south Asian cuisines use cashews for this unusual characteristic, rather than other nuts.

The shell of the cashew nut is toxic, which is why the shell is removed before it is sold to consumers.

Figure: *Cashew nuts, salted*

Cashew nuts are commonly used in Indian cuisine, whole for garnishing sweets or curries, or ground into a paste that forms a base of sauces for curries (e.g., *korma*), or some sweets (e.g., *kaju barfi*). It is also used in powdered form in the preparation of several Indian sweets and desserts. In Goan cuisine, both roasted and raw kernels are used whole for making curries and sweets.

The cashew nut can also be harvested in its tender form, when the shell has not hardened and is green in colour. The shell is soft and can be cut with a knife and the kernel extracted, but it is still corrosive at this stage, so gloves are required. The kernel can be soaked in turmeric water to get rid of the corrosive material before use. Cashew nuts are also used in Thai and Chinese cuisine, generally in whole form.

In the Philippines, cashew is a known product of Antipolo, and is eaten with *suman*. Pampanga also has a sweet dessert called *turrones de casuy*, which is cashew marzipan wrapped in white wafers.

In Indonesia, roasted and salted cashew nut is called *kacang mete* or *kacang mede*, while the cashew apple is called *jambu monyet* (literally means monkey rose apple).

In Mozambique, *bolo polana* is a cake prepared using powdered cashews and mashed potatoes as the main ingredients. This dessert is popular in South Africa, too.

South American countries have developed their own specialities. In Brazil, the cashew fruit juice is popular all across the country. In Panama, the cashew fruit is cooked with water and sugar for a prolonged time to make a sweet, brown, paste-like dessert called *dulce de marañón*. *Marañón* is one of the Spanish names for cashew.

Production

Nigeria was the world's largest producer of cashew nuts with shell in 2010. Cashew nut production trends have varied over the decades. African countries used to be the major producers before the 1980s; India became the largest producer in the 1990s, followed by Vietnam which became the largest producer in the mid-2000s. Since 2008, Nigeria has become the largest producer. Cashew nuts are produced in tropical countries because the tree is very frost sensitive; they have been adapted to various climatic regions around the world between the latitudes of 25°N and 25°S.

Peru reported the world's highest production yields for cashew nuts in 2010, at 5.27 metric tons per hectare, nearly nine times the world average. The traditional cashew tree is tall (up to 14 m) and takes three years from planting before it starts production, and eight years before economic harvests can begin. More recent breeds, such as the dwarf cashew trees, are up to 6 m tall, and start producing after the first year, with economic yields after three years. The cashew nut yields for the traditional tree are about 0.25 metric tons per hectare, in contrast to over a ton per hectare for the dwarf variety. Grafting and other modern tree management technologies are used to further improve and sustain cashew nut yields in commercial orchards.

Fluctuations in world market prices for cashew nuts have been a source of discontent for communities in Tanzania which grow the nut as a cash crop; reduced payments in April 2013 sparked serious rioting in Liwale District in the south of the country.

Nutrition

The fats and oils in cashew nuts are 54% monounsaturated fat (18:1), 18% polyunsaturated fat (18:2), and 16% saturated fat (9% palmitic acid (16:0) and 7% stearic acid (18:0)). Cashews, as with other tree nuts, are a good source of antioxidants. Alkyl phenols, in particular, are abundant in cashews. Cashews are also a source of dietary trace minerals copper, manganese, magnesium and phosphorus.

Allergy

For some people, cashews, like other tree nuts, can lead to complications or allergic reactions. Cashews contain gastric and intestinal soluble oxalates, albeit less than some other tree nuts; people with a tendency to form kidney stones may need moderation and medical guidance. Allergies to tree nuts such as cashews can be of severe nature to some people. These allergic reactions can be life-threatening or even fatal; prompt medical attention is necessary if tree nut allergy reaction is observed. These allergies are triggered by the proteins found in tree nuts, and cooking often does not remove or change these proteins. Reactions to cashew and other tree nuts can also occur as a consequence of hidden nut ingredients or traces of nuts that may inadvertently be introduced during food processing, handling or manufacturing. Many nations require food label warning if the food may get inadvertent exposure to tree nuts such as cashews.

In some people, cashew nut allergy may be a different form, namely birch pollen allergy. This is usually a minor form. Symptoms are confined largely to the mouth.

Cashew Oil

Cashew oil is a dark yellow oil for cooking or salad dressing pressed from cashew nuts (typically broken chunks created during processing). This may be produced from a single cold pressing.

Cashew Shell Oil

Cashew nutshell liquid (CNSL) or cashew shell oil (CAS registry number 8007-24-7) is a natural resin found in the honeycomb structure of the cashew nutshell and is a byproduct of processing cashew nuts. It is a raw material of multiple uses in developing drugs, antioxidants, fungicides, etc. It is used in tropical folk medicine and for anti-termite treatment of timber. Its composition varies depending on how it is processed.

- Cold, solvent extracted CNSL is mostly composed of anacardic acids (70%), cardol (18%) and cardanol (5%).
- Heating CNSL decarboxylates the anacardic acids, producing a technical grade of CNSL that is rich in cardanol. Distillation of this material gives distilled, technical CNSL containing 78% cardanol and 8% cardol (cardol has one more hydroxyl group than cardanol). This process also reduces the degree of thermal polymerization of the un-saturated alkyl-phenols present in cashew shell nut liquid.
- Anacardic acid is also used in the chemical industry for the production of cardanol, which is used for resins, coatings, and frictional materials.

These substances are dermatogenic, like the oils of the poison ivy, and present danger during manual cashew processing.

This natural oil phenol has been found to have interesting chemical structural features which enable a range of chemical modifications to create a wide spectrum of bio-based monomers capitalising on the chemically versatile construct, containing three different functional groups, the aromatic ring, the hydroxyl group and the double bonds in the flanking alkyl chain. These can be split into key groups, used as polyols, which have recently seen a dramatic increase in demand for their bio-based origin and key chemical attributes such as high reactivity, range of functionalities, reduction in blowing agents and naturally occurring fire retardant properties in the field of ridged polyurethanes aided by their inherent phenolic structure and larger number of reactive units per unit mass.

CNSL based Novolac is another versatile industrial monomer deriving from cardanol typically used as a reticulating agent for epoxy matrices in composite applications providing good thermal and mechanical properties to the final composite material.

Further examples of applications which are cashew shell nut liquid derived materials are being evaluated, are in the fields of chemical intermediates, additives, stabilizers, lubricants, diesel engine fuel alternatives, poor point dispersants, anti-oxidants, and anticorrosive paints.

Abrasives and friction dusts have also been developed from Residol, the residue byproduct of this synthesis process.

During World War II cashew shell oil was used as a sabotage device by the OSS (Office of Strategic Services) for destroying the engines of German vehicles when poured into the crankcase of engines.

Other uses of cashew shell oil have been explored, including as an additive to brake fluid, to reduce brake fade and brake lining wear.

Composite Technical Services (Kettering, Ohio, USA) has researched the use of cashew shell oil as a resin for carbon composite products.

Cashew Apple

Figure: *Cashew apple with nut*

The cashew apple, also called cashew fruit, is the fleshy part of the cashew fruit that is attached to the cashew nut. The top end of the cashew apple is attached to the stem that comes off the tree. The bottom end of the cashew apple attaches to the cashew nut, which is encased in a shell. In botanical terms, the cashew apple is an accessory fruit that grows on the cashew seed (which is the nut).

The cashew apple is a soft fruit, rich in nutrients, and contains five times more vitamin C than an orange. It is eaten fresh, cooked in curries, or fermented into vinegar, as well as an alcoholic drink. It is also used to make preserves, chutneys, and jams in some countries such as India and Brazil. In many countries, particularly in South America, the cashew apple is used to flavour drinks, both alcoholic and nonalcoholic. In Brazil, it is a popular fruit flavour for the national drink, the *caipirinha*.

In much of South America, people regard the cashew apple as the delicacy, rather than the nut kernel which is popular elsewhere. In fact, in many South American countries, the cashew apple is more popular as a food than is the cashew nut. A large reason for this is simply the availability of cashew apples. They tend to be popular where they are readily available.

Cashew nuts are more popular than cashew apples in many parts of the world—regions that do not grow cashews—because the fruit, unlike the nut, is difficult to transport to these places. Unlike cashew nuts, cashew apples are extremely soft and easily bruised in shipment. For this reason, cashew juice and cashew juice concentrate are often shipped to these nonlocal countries instead of the fresh fruit.

Cashew apples have a sweet but astringent taste. This astringency has been traced to the waxy layer on the skin that contains a chemical, urushiol, which can cause minor skin irritation to areas that have had contact with it. It is almost identical to the astringency caused by the skin of a mango, which also contains urushiol.

The astringency from mango skin can be mildly tasted in the flesh of mango fruit, just as the astringency of cashew apple skin can be mildly tasted in the flesh of cashew apples. In cultures that consume cashew apples, this astringency is sometimes removed by steaming the fruit for five minutes before washing it in cold water; alternatively, boiling the fruit in salt water for five minutes or soaking it in gelatin solution also reduces the astringency.

When mixed in drinks or used as a flavoring, the astringency becomes highly diluted and typically causes no irritation to those without urushiol allergies.

Alcohol

Figure: *Young cashew nuts*

In Goa, the cashew apple (the accessory fruit) is mashed, the juice is extracted and kept for fermentation for a few days. Fermented juice then undergoes a double distillation process. The resulting beverage is called *feni* or fenny. *Feni* is about 40-42% alcohol. The single-distilled version is called *urrac*, which is about 15% alcohol.

In the southern region of Mtwara, Tanzania, the cashew apple (*bibo* in Swahili) is dried and saved. Later it is reconstituted with water and fermented, then distilled to make a strong liquor often referred to by the generic name, *gongo*.

In Mozambique, cashew farmers commonly make a strong liquor from the cashew apple, *agua ardente* (burning water).

According to *An Account of the Island of Ceylon* by Robert Percival an alcohol had been distilled in the early 20th century from the juice of the fruit, and had been manufactured in the West Indies. Apparently, the Dutch considered it superior to brandy as a liqueur.

Medicinal Uses

Many parts of the plant are used in the traditional medicine of the Patamona of Guyana. They grind the seeds into a poultice for treating snakebites, apply nut oil to cracked heels or as an antifungal agent, and use the fruits, bark, and leaves for many other purposes including anti-fungal activity, for sores and rashes, or as an antipyretic, and for antidiarrheal applications. The leaf extracts with petroleum ether and ethanol inhibited growth of several species of bacteria and fungi.

Chemicals identified in cashew shell oil have been assayed against *Streptococcus mutans*, a bacterium responsible for many dental cavities, and found to have activity *in vitro* against this and other Gram positive bacteria.

Palmyrah

Borassus (Palmyra Palm) is a genus of six species of fan palms, native to tropical regions of Africa, Asia and New Guinea. They are tall palms, capable of growing up to 30 m high (98 ft). The leaves are long, fan-shaped, 2 to 3 m in length. The flowers are small, in densely clustered spikes, followed by large, brown, roundish fruits.

Species

- *Borassus aethiopum* - African Palmyra Palm (and other names) (tropical Africa)
- *Borassus akeassii* - Ake Assi's Palmyra Palm (West Africa)
- *Borassus flabellifer* - Asian Palmyra palm/Lontar palm/Doub palm/Sea Coconut (southern Asia and southeast Asia)
- *Borassus heineanus* - New Guinea Palmyra Palm (New Guinea)
- *Borassus madagascariensis* - Madagascar Palmyra Palm (Madagascar)
- *Borassus sambiranensis* - Sambirano Palmyra Palm (Madagascar)

Cultivation and Uses

Figure: *The main entrance of Angkor Wat to the temple proper, seen from the eastern end of the Nâga causeway and Asian Palmyra Palm*

Palmyra palms are economically useful, and widely cultivated in tropical regions. The palmyra palm has long been one of the most important trees of Cambodia and India, where it has over 800 uses. The leaves are used for thatching, mats, baskets, fans, hats, umbrellas, and as writing material.

In Cambodia, the tree is a national flora symbol/emblem that is seen growing around Angkor Wat. The sugar palm can live over 100 years.

In ancient Indonesia and ancient India, the leaves were used as paper to write on, as a kind of papyrus. In India, leaves of suitable size, shape and texture, and sufficient maturity are chosen. They are then preserved by boiling in salt water with turmeric powder. The leaves are then dried; when they are dry enough, the faces of the leaves are polished with pumice stone. Then they are cut in the proper size. A hole is cut out in one corner. Each leaf will have four pages. The writing is done with a stylus. The writing is of a very cursive and interconnected style. The leaves are then tied up as sheaves.

The stalks are used to make fences and to make a strong, wiry fibre suitable for cordage and brushes. The black timber is hard, heavy, and durable and is highly valued for construction, such as for wharf pilings.

***Figure:** Ake Assi's Palmyra Palm (*Borassus akeassii*) fruit*

***Figure:** Jelly like seeds of Palmyra Palm (*Borassus akeassii*) fruit*

The tree also yields many types of food. The young plants are cooked as a vegetable or roasted and pounded to make meal. The fruits are eaten roasted or raw, and the young, jellylike seeds are also eaten. A sugary sap, called toddy, can be obtained from the young inflorescence, either male or female ones. toddy is (called "kallu (2M2A)" in Telugu. The toddy is fermented to make a beverage called arrack, or it is concentrated to a crude sugar called jaggery/palm sugar. It is called *Gula Jawa* (Javanese sugar) in Indonesia and is widely used in the Javanese cuisine. The roots can be dried to form Odiyal, a hard chewable snack. In addition, the tree sap is taken as a laxative, and medicinal values have been ascribed to other parts of the plant.

In Tamil Culture

The Palmyra tree is the official tree of Tamil Nadu. In Tamil culture it is called *karpaha*,"Nungu" "celestial tree", and is highly respected because all its parts can be used. The recently germinated seeds form fleshy sprouts below the surface which can be boiled and eaten as a fibrous, nutritious food. The germinated seed's hard shell is also cut open to take out the crunchy kernel which tastes like a water chestnut but is sweeter. The ripe fibrous outer layer of the fruits is edible after boiling or roasting. When the fruit is tender, the kernel inside the hard shell is an edible jelly that is refreshing and rich in minerals. When the crown of the tree from which the leaves sprout is cut we get an edible cake. In ancient times, dried palm leaves were used to write manuscripts.

Cardamom

Cardamom refers to several plants of the similar genera *Elettaria* and *Amomum* in the ginger family Zingiberaceae. Both genera are native to India, Pakistan, Nepal, and Bhutan; they are recognised by their small seed pods, triangular in cross-section and spindle-shaped, with a thin, papery, outer shell and small black seeds. Guatemala is the biggest producer and exporter of cardamom in the world, followed by India. Some other countries, such as Sri Lanka, have also begun to cultivate it. *Elettaria* pods are light green, while *Amomum* pods are larger and dark brown.

It is the world's third-most expensive spice by weight, outstripped in market value only by saffron and vanilla.

Types and Distribution

The two main genera of the ginger family named as forms of cardamom are distributed as follows:

- *Elettaria* (commonly called cardamom, green cardamom, or true cardamom) is distributed from India to Malaysia.
- *Amomum* (commonly known as black cardamom, brown cardamom, *kravan*, Java cardamom, Bengal cardamom, Siamese cardamom, white cardamom, or red cardamom) is distributed mainly in Asia and Australia.

The two types of cardamom, were distinguished in the fourth century BCE by the Greek father of botany, Theophrastus. Some of his informants told him these varieties came to Greece from the land of the Medes in northern Persia, while others were aware it came originally from India.

Figure: *Green and black cardamom*

Ecology

Elettaria cardamomum is used as a food plant by the larvae of the moth *Endoclita hosei.*

Varieties

The three natural varieties of green cardamom plants are:

- Malabar (Nadan/native), as the name suggests, is the native variety of Kerala. These plants have panicles which grow horizontally along the ground.
- Mysore, as the name suggests, is a native variety of Karnataka. These plants have panicles which grow vertically upwards. The Mysore variety has, however, declined in the past few decades owing to the emergence of the more resistant and better yielding 'Green Gold' variety, and which is the most common form of cardamom harvested in Kerala.
- Vazhuka is a naturally occurring hybrid between Malabar and Mysore varieties, and the panicles grow neither vertically nor horizontally, but in between.

Recently, a few planters isolated high-yielding plants and started multiplying them on a large scale. The most popular high-yielding variety is 'Njallani'. 'Njallani', also known as *rup-ree-t*, is a unique high-yielding cardamom variety developed by an Indian farmer, Sebastian Joseph, at Kattappana in the South Indian state of Kerala. K. J. Baby of Idukki District, Kerala, has developed a purely white-flowered variety of Vazhuka type green cardamom having higher yield than 'Njallani'. The variety has high adaptability to different shade conditions and can also be grown in waterlogged areas.

Figure: *Cardamom plant*

Uses

Both forms of cardamom are used as flavorings and cooking spices in both food and drink, and as a medicine. *E. cardamomum* (green cardamom) is used as a spice, a masticatory, and in medicine; it is also smoked, sometimes.

Food and Drink

Cardamom has a strong, unique taste, with an intensely aromatic, resinous fragrance. Black cardamom has a distinctly more smokey, though not bitter, aroma, with a coolness some consider similar to mint.

Green cardamom is one of the most expensive spices by weight, but little is needed to impart flavour. It is best stored in pod form because once the seeds are exposed or ground, they quickly lose their flavour. However, high-quality ground cardamom seed is often more readily (and cheaply) available and is an acceptable substitute. Grinding the pods and seeds together lowers both the quality and the price. For recipes requiring whole cardamom pods, a generally accepted equivalent is 10 pods equals 1½ teaspoons of ground cardamom.

***Figure:** Spice shop in Sri Lanka*

It is a common ingredient in Indian cooking and is often used in baking in Nordic countries, such as in the Finnish sweet bread *pulla* or in the Scandinavian Christmas bread *Julekake*. In the Middle East, green cardamom powder is used as a spice for sweet dishes, as well as traditional flavouring in coffee and tea. Cardamom is used to some extent in savoury dishes. In some Middle Eastern countries, coffee and cardamom are often ground in a wooden mortar, a *mihbaj*, and cooked together in a skillet, a *mehmas*, over wood or gas, to produce mixtures as much as 40% cardamom.

In South Asia, green cardamom is often used in traditional Indian sweets and in *masala chai* (spiced tea). Black cardamom is sometimes used in *garam masala* for curries. It is occasionally used as a garnish in basmati rice and other dishes. It is often referred to as fat cardamom due to its size. Individual seeds are sometimes chewed and used in much the same way as chewing gum. It is even used by confectionery giant Wrigley; its Eclipse Breeze Exotic Mint packaging indicates it contains "cardamom to neutralize the toughest breath odors". It has been known to be used in making gin and herbal teas.

Traditional Medicine

Green cardamom is broadly used in South Asia to treat infections in teeth and gums, to prevent and treat throat troubles, congestion of the lungs and pulmonary tuberculosis, inflammation of eyelids, and digestive disorders. It also is used to break up kidney and gall stones, and was reportedly used as an antidote for both snake and scorpion venoms. *Amomum* is used as a spice and as an ingredient in traditional medicine in systems of the traditional Chinese medicine in China, in Ayurveda in India, Pakistan, Japan, Korea, and Vietnam. Among other species, varieties, and cultivars, *Amomum villosum* cultivated in China, Laos, and Vietnam is used in traditional Chinese medicine to treat stomach problems, constipation, dysentery, and other digestion problems. *Tsaoko* cardamom, *Amomum tsao-ko,* is cultivated in Yunnan and northwest Vietnam, both for medicinal purposes and as a spice.

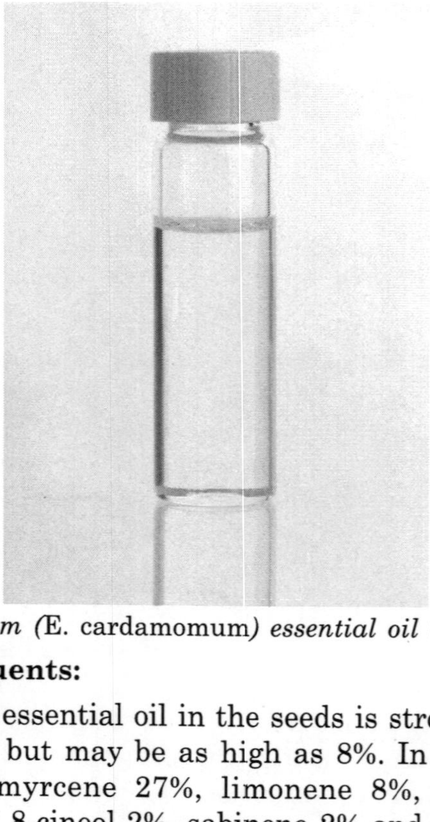

Figure: *Cardamom (*E. cardamomum*) essential oil in clear glass vial*

Main constituents:

The content of essential oil in the seeds is strongly dependent on storage conditions, but may be as high as 8%. In the oil were found á-terpineol 45%, myrcene 27%, limonene 8%, menthone 6%, â-phellandrene 3%, 1,8-cineol 2%, sabinene 2% and heptane 2%. Other sources report 1,8-cineol (20 to 50%), á-terpenylacetate (30%), sabinene, limonene (2 to 14%), and borneol.

In the seeds of round cardamom from Jawa (*A. kepulaga*), the content of essential oil is lower (2 to 4%), and the oil contains mainly 1,8 cineol (up to 70%) plus â-pinene (16%); further-more, á-pinene, á-terpineol and humulene were found.

World Production

Guatemala is the largest producer of cardamom in the world, with an average annual yield between 25,000 to 29,000 metric tons. India is the second producer worldwide (formerly the largest), generating around 15,000 mt annually. Cardamom was first introduced to Guatemala in 1914.

Increased demand since the 1980s, principally from China, for both *A. villosum* and *A. tsao-ko*, has provided a key source of income for poor farmers living at higher altitudes in localized areas of China, Laos, and Vietnam, people typically isolated from many other markets. Nepal was previously the world's largest producer of large cardamom.

According to estimates of the Asociación de Cardamomeros de Guatemala (Cardegua), the harvest of 2012 will reach to about 29,000 mt, 12% more than in 2011, when it was 26,000 mt.

Pistachio

The pistachio, *Pistacia vera,* a member of the cashew family, is a small tree originally from Central Asia and the Middle East. Pistachio trees can be found in regions of Iran, Syria, Lebanon, Turkey, Greece, Tunisia, Kyrgyzstan, Tajikistan, Turkmenistan, India, Pakistan, Egypt, Italy (Sicily), Uzbekistan, Afghanistan (especially in the provinces of Samangan and Badghis), and the United States, specifically in California. The tree produces seeds that are widely consumed as food.

However, it is only native to the eastern Mediterranean (Cyprus and Turkey to Israel and Syria) and Central Asia (Turkmenistan, Uzbekistan, Afghanistan, Tajikistan, and Kyrgyzstan).

Pistacia vera often is confused with other species in the genus *Pistacia* that are also known as pistachio. These species can be distinguished from *P. vera* by their geographic distributions (in the wild), and their seeds which are much smaller and have a shell that is not hard.

Botany

Habitat: Pistachio is a desert plant, and is highly tolerant of saline soil. It has been reported to grow well when irrigated with water having 3,000–4,000 ppm of soluble salts. Pistachio trees are fairly hardy in the right conditions, and can survive temperatures ranging between "10 °C (14 °F) in winter and 48 °C (118 °F) in summer.

They need a sunny position and well-drained soil. Pistachio trees do poorly in conditions of high humidity, and are susceptible to root rot in winter if they get too much water and the soil is not sufficiently free-draining. Long, hot summers are required for proper ripening of the fruit.

The Jylgyndy Forest Reserve, a preserve protecting the native habitat of *Pistacia vera* groves, is located in the Nooken District of Jalal-Abad Province of Kyrgyzstan.

Figure: *Leaves of a pistachio tree in Syria.*

Characteristics

Figure: *Pistachios in shell*

The bush grows up to 10 metres (33 ft) tall. It has deciduous pinnate leaves 10–20 centimetres (4–8 inches) long. The plants are dioecious, with separate male and female trees. The flowers are apetalous and unisexual, and borne in panicles.

The fruit is a drupe, containing an elongated seed, which is the edible portion. The seed, commonly thought of as a nut, is a culinary nut, not a botanical nut. The fruit has a hard, whitish exterior shell. The seed has a mauvish skin and light green flesh, with a distinctive flavour. When the fruit ripens, the shell changes from green to an autumnal yellow/red, and abruptly splits part way open. This is known as dehiscence, and happens with an audible pop. The splitting open is a trait that has been selected by humans. Commercial cultivars vary in how consistently they split open.

Each pistachio tree averages around 50 kilograms (110 lb) of seeds, or around 50,000, every two years.

The shell of the pistachio is naturally a beige colour, but it is sometimes dyed red or green in commercial pistachios. Originally, dye was applied by importers to hide stains on the shells caused when the seeds were picked by hand.

Most pistachios are now picked by machine and the shells remain unstained, making dyeing unnecessary except to meet ingrained consumer expectations. Roasted pistachio seeds can be artificially turned red if they are marinated prior to roasting in a salt and strawberry marinade, or salt and citrus salts.

Like other members of the Anacardiaceae family (which includes poison ivy, sumac, mango, and cashew), pistachios contain urushiol, an irritant that can cause allergic reactions.

Cultivation

Iran, the United States and Turkey are the major producers of pistachios. The trees are planted in orchards, and take approximately seven to ten years to reach significant production. Production is alternate bearing or biennial bearing, meaning the harvest is heavier in alternate years. Peak production is reached at approximately 20 years. Trees are usually pruned to size to make the harvest easier.

One male tree produces enough pollen for eight to twelve drupe-bearing females. Harvesting in the United States and in Greece is often accomplished by using shaking equipment to shake the drupes off the tree. After hulling and drying, pistachios are sorted according to open mouth and closed mouth shell. Sun drying has been found to be the best method of drying.

Then they are roasted or processed by special machines to produce pistachio kernels.

Pistachio trees are vulnerable to a wide variety of diseases. Among these is infection by the fungus *Botryosphaeria*, which causes panicle and shoot blight (*i.e.,* kills flowers and young shoots), and can damage entire pistachio orchards.

In California, almost all female pistachio trees are the cultivar "Kerman". A scion from a mature female Kerman is grafted onto a one-year-old rootstock. Male pistachios may be a different variety.

In 2013, the City Council of Avenal, California self-proclaimed Avenal to be the Pistachio Capital of the World, reflecting the importance of the pistachio industry to the city's economy.

In Greece, the cultivated type of pistachios is different. It has an almost-white shell, a sweet taste, a red-green kernel and a little bit more close mouth shell than "Kerman" variety. Most of the production in Greece comes from the island of Aegina and the region of Thessaly - Almyros.

Bulk container shipments of pistachio kernels are prone to self-heating and spontaneous combustion because of their high fat and low water content.

Consumption

Figure: *Pistachio Turkish delight*

The kernels are often eaten whole, either fresh or roasted and salted, and are also used in pistachio ice cream, kulfi, spumoni, historically in Neopolitan ice cream, pistachio butter, pistachio paste and confections such as baklava, pistachio chocolate, pistachio halva, pistachio lokum or biscotti and cold cuts such as mortadella. Americans make pistachio salad, which includes fresh pistachios or pistachio pudding, whipped cream, and canned fruit.

In July 2003, the Food and Drug Administration (FDA) approved the first qualified health claim specific to seeds lowering the risk of heart disease: "Scientific evidence suggests but does not prove that eating 1.5 ounces (42.5 g) per day of most nuts, such as pistachios, as part of a diet low in saturated fat and cholesterol may reduce the risk of heart disease".

China is the top pistachio consumer worldwide, with annual consumption of 80,000 tons, while the United States consumes 45,000 tons. Russia (with consumption of 15,000 tons) and India (with consumption of 10,000 tons) are in the third and fourth places.

Recycling the Shells

Empty pistachio shells can be recycled in several ways. If unsalted, the shells need neither washing nor drying before reuse but washing is simple if this is not the case. Practical uses include as a fire starter, just as crumpled paper is used as kindling; to line the bottom of pots containing houseplants, for drainage and retention of soil for up to two years; as a mulch for shrubs and plants that require acid soils; as a medium for orchids; and as an addition to a compost pile designed for wood items that take longer to decompose than leafy materials, taking up to a year for pistachio shells to decompose unless soil is added to the mix. Shells from salted pistachios can also be placed around the base of plants to deter slugs and snails. Craft uses for the shells include Christmas tree ornaments, jewellery, mosaics, and rattles. Research indicates that pistachio shells may be helpful in cleaning up pollution created by mercury emissions.

3

Arable Crops

In geography and agriculture, arable land is land ploughed or tilled regularly, generally under a system of crop rotation. Permanent crops that occupy the land for a number of years, and don't need replanting after each annual harvest — like coffee, rubber, flowering shrubs, fruit, nut trees and vines — are not counted as existing on arable land, but as existing on permanent cropland.

Permanent pastures and meadows used for grazing, land mowed for hay or silage not included in a crop rotation scheme, and abandoned land resulting from shifting cultivation is also not counted as arable, along with lands with built-on and barren areas, forests and woodlands. Although constrained by land mass and topography, the amount of arable land, both regionally and globally, fluctuates due to human and climatic factors such as irrigation, deforestation, desertification, terracing, land reclamation and urban sprawl. In 2008, the world's arable land amounted to 1,386 M ha, out of a total 4,883 M ha land used for agriculture.

Tobacco

Tobacco is a plant within the genus Nicotiana of the Solanaceae (nightshade) family. While there are more than 70 species of tobacco, the chief commercial crop is *N. tabacum*. The more potent species *N. rustica* is also widely used around the world. Dried tobacco leaves are mainly smoked in cigarettes, cigars, pipe tobacco and flavored shisha tobacco. They are also consumed as snuff, chewing tobacco and dipping tobacco.

Traditional Use

Figure: *The earliest depiction of a European man smoking, from* Tabacco *by Anthony Chute.*

Tobacco had already long been used in the Americas, with some cultivation sites in Mexico dating back to 1400–1000 B.C. Many Native American tribes have traditionally grown and used tobacco as an entheogen. Eastern North American tribes carried large amounts of tobacco in pouches as a readily accepted trade item, and often smoked it in peace pipes, either in defined sacred ceremonies, or to seal a bargain. They smoked it at such occasions in all stages of life, even in childhood. It was believed that tobacco is a gift from the Creator, and that the exhaled tobacco smoke carries one's thoughts and prayers to heaven.

Popularization

Figure: *An Illustration from Frederick William Fairholt's* Tobacco, its History and Association, *1859.*

Following the arrival of the Europeans, tobacco became increasingly popular as a trade item. Before the development of lighter Virginia and White Burley strains of tobacco, the smoke was too harsh to be inhaled. small quantities at a time, using a pipe like the midwakh or kiseru or smoking newly invented waterpipes such as the bong or the hookah. Inhaling smoke was already common in India and China through the consumption of cannabis and opium millennia before.

Tobacco fostered the economy for the southern United States until it was replaced by cotton. Following the American civil war, a change in demand and a change in labour force allowed inventor James Bonsack to create a machine that automated cigarette production.This increase in production allowed tremendous growth in the tobacco industry until the scientific revelations of the mid-20th century.

Biology

Nicotiana: There are many species of tobacco in the genus of herbs *Nicotiana*. It is part of the nightshade family (Solanaceae) indigenous to North and South America, Australia, South West Africa and the South Pacific.

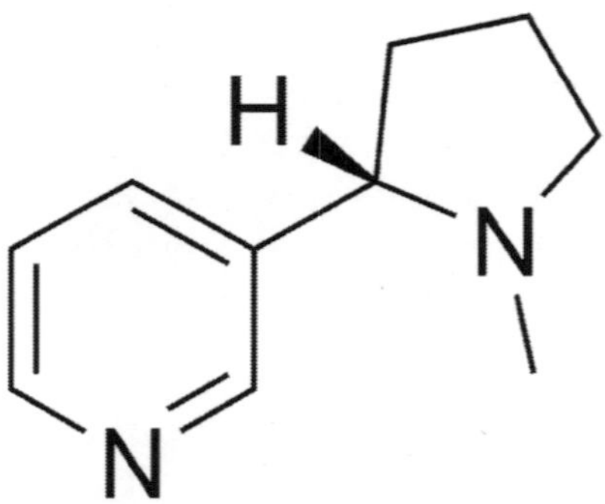

Figure: *Nicotine is the compound responsible for the addictive nature of Tobacco use.*

Many plants contain nicotine, a powerful neurotoxin to insects. However, tobaccos contain a higher concentration of nicotine than most other plants. Unlike many other Solanaceae, they do not contain tropane alkaloids, which are often poisonous to humans and other animals.

Despite containing enough nicotine and other compounds such as germacrene and anabasine and other piperidine alkaloids (varying between species) to deter most herbivores, a number of such animals have evolved the ability to feed on *Nicotiana* species without being harmed. Nonetheless, tobacco is unpalatable to many species, and accordingly some tobacco plants (chiefly tree tobacco, *N. glauca*) have become established as invasive weeds in some places.

Types

There are a number of types of tobacco including, but are not limited to:

- *Aromatic fire-cured* is cured by smoke from open fires. In the United States, it is grown in northern middle Tennessee, central Kentucky and in Virginia. Fire-cured tobacco grown in Kentucky and Tennessee are used in some chewing tobaccos, moist snuff, some cigarettes, and as a condiment in pipe tobacco blends. Another fire-cured tobacco is Latakia, which is produced from oriental varieties of *N. tabacum*. The leaves are cured and smoked over smoldering fires of local hardwoods and aromatic shrubs in Cyprus and Syria.
- *Brightleaf tobacco*, Brightleaf is commonly known as "Virginia tobacco", often regardless of the state where they are planted. Prior to the American Civil War, most tobacco grown in the US was fire-cured dark-leaf. This type of tobacco was planted in fertile lowlands, used a robust variety of leaf, and was either fire cured or air cured. Most Canadian cigarettes are made from 100% pure Virginia tobacco.
- *Burley tobacco*, is an air-cured tobacco used primarily for cigarette production. In the U.S., burley tobacco plants are started from palletized seeds placed in polystyrene trays floated on a bed of fertilized water in March or April.
- *Cavendish* is more a process of curing and a method of cutting tobacco than a type. The processing and the cut are used to bring out the natural sweet taste in the tobacco. Cavendish can be produced from any tobacco type, but is usually one of, or a blend of Kentucky, Virginia, and burley, and is most commonly used for pipe tobacco and cigars.
- *Criollo tobacco* is a type of tobacco, primarily used in the making of cigars. It was, by most accounts, one of the original Cuban tobaccos that emerged around the time of Columbus.
- *Dokha*, is a tobacco originally grown in Iran, mixed with leaves, bark, and herbs for smoking in a midwakh.
- *Turkish tobacco*, is a sun-cured, highly aromatic, small-leafed variety (*Nicotiana tabacum*) that is grown in Turkey, Greece, Bulgaria, and Macedonia. Originally grown in regions historically part of the Ottoman Empire, it is also known as "oriental". Many of the early brands of cigarettes were made mostly or entirely of Turkish tobacco; today, its main use is in blends of pipe and

especially cigarette tobacco (a typical American cigarette is a blend of bright Virginia, burley and Turkish).

- *Perique*, a farmer called Pierre Chenet is credited with first turning this local tobacco into the Perique in 1824 through the technique of pressure-fermentation. Considered the truffle of pipe tobaccos, it is used as a component in many blended pipe tobaccos, but is too strong to be smoked pure. At one time, the freshly moist Perique was also chewed, but none is now sold for this purpose. It is typically blended with pure Virginia to lend spice, strength, and coolness to the blend.
- *Shade tobacco*, is cultivated in Connecticut and Massachusetts. Early Connecticut colonists acquired from the Native Americans the habit of smoking tobacco in pipes, and began cultivating the plant commercially, even though the Puritans referred to it as the "evil weed". The Connecticut shade industry has weathered some major catastrophes, including a devastating hailstorm in 1929, and an epidemic of brown spot fungus in 2000, but is now in danger of disappearing altogether, given the increase in the value of land.
- *White burley*, in 1865, George Webb of Brown County, Ohio planted red burley seeds he had purchased, and found that a few of the seedlings had a whitish, sickly look. The air-cured leaf was found to be more mild than other types of tobacco.
- *Wild tobacco*, is native to the southwestern United States, Mexico, and parts of South America. Its botanical name is *Nicotiana rustica*.
- *Y1* is a strain of tobacco cross-bred by Brown & Williamson in the 1970s to obtain an unusually high nicotine content. In the 1990s, the United States Food and Drug Administration (FDA) used it as evidence that tobacco companies were intentionally manipulating the nicotine content of cigarettes.

Cultivation

Tobacco is cultivated similarly to other agricultural products. Seeds were at first quickly scattered onto the soil. However, young plants came under increasing attack from flea beetles (*Epitrix cucumeris* or *Epitrix pubescens*), which caused destruction of half the tobacco crops in United States in 1876. By 1890, successful experiments were conducted that placed the plant in a frame covered by thin cotton fabric. Today, tobacco is sown in cold frames or hotbeds, as their germination is activated by light.

In the United States, tobacco is often fertilized with the mineral apatite, which partially starves the plant of nitrogen, to produce a more desired flavour.

Figure: *Tobacco plants growing in a field in Intercourse, Pennsylvania.*

After the plants are about eight inches tall, they are transplanted into the fields. Farmers used to have to wait for rainy weather to plant. A hole is created in the tilled earth with a tobacco peg, either a curved wooden tool or deer antler. After making two holes to the right and left, the planter would move forward two feet, select plants from his/her bag, and repeat. Various mechanical tobacco planters like Bemis, New Idea Setter, and New Holland Transplanter were invented in the late 19th and 20th centuries to automate the process: making the hole, watering it, guiding the plant in — all in one motion.

Tobacco is cultivated annually, and can be harvested in several ways. In the oldest method still used today, the entire plant is harvested at once by cutting off the stalk at the ground with a tobacco knife. It is then speared onto sticks, four to six plants a stick and hung in a curing barn. In the 19th century, bright tobacco began to be harvested by pulling individual leaves off the stalk as they ripened. The leaves ripen from the ground upwards, so a field of tobacco harvested in this manner will involve the serial harvest of a number of "primings," beginning with the volado leaves near the ground, working to the seco leaves in the middle of the plant, and finishing with the potent ligero leaves at the top. Before this the crop needs to be topped when the pink flowers develop. Topping always refers to the removal of the tobacco flower before the leaves are systematically removed and, eventually, entirely harvested. As the industrial revolution took hold, harvesting wagons used to transport leaves were equipped with man-powered stringers, an apparatus that used twine to attach leaves to a pole. In modern times, large fields are harvested mechanically, although topping the flower and in some cases the plucking of immature leaves is still done by hand.

Consumption

Tobacco is consumed in many forms and through a number of different methods. Below are examples including, but not limited to, such forms and usage.

- Beedi are thin, often flavoured cigarettes from India made of tobacco wrapped in a tendu leaf, and secured with coloured thread at one end.
- Chewing tobacco is the oldest way of consuming tobacco leaves. It is consumed orally, in two forms: through sweetened strands, or in a shredded form. When consuming the long sweetened strands, the tobacco is lightly chewed and compacted into a ball. When consuming the shredded tobacco, small amounts are placed at the bottom lip, between the gum and the teeth, where it is gently compacted, thus it can often be called *dipping tobacco*. Both methods stimulate the saliva glands, which led to the development of the spittoon.
- Cigars are tightly rolled bundles of dried and fermented tobacco, which is ignited so its smoke may be drawn into the smoker's mouth.
- Cigarettes are a product consumed through inhalation of smoke and manufactured from cured and finely cut tobacco leaves and reconstituted tobacco, often combined with other additives, then rolled into a paper cylinder.
- Creamy snuffs are tobacco paste, consisting of tobacco, clove oil, glycerin, spearmint, menthol, and camphor, and sold in a toothpaste tube. It is marketed mainly to women in India, and is known by the brand names Ipco (made by Asha Industries), Denobac, Tona, Ganesh. It is locally known as "mishri" in some parts of Maharashtra.
- Dipping tobaccos are a form of smokeless tobacco. Dip is occasionally referred to as "chew", and because of this, it is

commonly confused with chewing tobacco, which encompasses a wider range of products. A small clump of dip is 'pinched' out of the tin and placed between the lower or upper lip and gums.

- Gutka is a preparation of crushed betel nut, tobacco, and sweet or savory flavorings. It is manufactured in India and exported to a few other countries. A mild stimulant, it is sold across India in small, individual-size packets.
- Hookah is a single or multi-stemmed (often glass-based) water pipe for smoking. Hookahs were first used in India and Persia, the hookah has gained immense popularity, especially in the Middle East. A hookah operates by water filtration and indirect heat. It can be used for smoking herbal fruits or moassel, a mixture of tobacco, flavouring and honey or glycerin.
- Kreteks are cigarettes made with a complex blend of tobacco, cloves and a flavoring "sauce". It was first introduced in the 1880s in Kudus, Java, to deliver the medicinal eugenol of cloves to the lungs.
- Roll-Your-Own, often called 'rollies' or 'roll-ups', are relatively popular in some European countries. These are prepared from loose tobacco, cigarette papers and filters all bought separately. They are usually cheaper to make.
- Pipe smoking typically consists of a small chamber (the bowl) for the combustion of the tobacco to be smoked and a thin stem (shank) that ends in a mouthpiece (the bit). Shredded pieces of tobacco are placed into the chamber and ignited.
- Snuff is a ground smokeless tobacco product, inhaled or "snuffed" through the nose.
- Snus is a steam-cured moist powder tobacco product that is not fermented, and does not induce salivation. It is consumed by placing it in the mouth against the gums for an extended period of time. It is a form of snuff used in a manner similar to American dipping tobacco, but does not require regular spitting.
- Topical tobacco paste is sometimes usedas a treatment for wasp, hornet, fire ant, scorpion, and bee stings. An amount equivalent to the contents of a cigarette is mashed in a cup with about a 0.5 to 1 teaspoon of water to make a paste that is then applied to the affected area.
- Tobacco water is a traditional organic insecticide used in domestic gardening. Tobacco dust can be used similarly. It is produced by boiling strong tobacco in water, or by steeping the tobacco

in water for a longer period. When cooled, the mixture can be applied as a spray, or 'painted' on to the leaves of garden plants, where it kills insects. Tobacco is however banned from use as pesticide in certified organic production.

Global Production

Trends: Production of tobacco leaf increased by 40% between 1971, during which 4.2 million tons of leaf were produced, and 1997, during which 5.9 million tons of leaf were produced. According to the Food and Agriculture organization of the UN, tobacco leaf production was expected to hit 7.1 million tons by 2010. This number is a bit lower than the record high production of 1992, during which 7.5 million tons of leaf were produced. The production growth was almost entirely due to increased productivity by developing nations, where production increased by 128%. During that same time period, production in developing countries actually decreased. China's increase in tobacco production was the single biggest factor in the increase in world production. China's share of the world market increased from 17% in 1971 to 47% in 1997. This growth can be partially explained by the existence of a high import tariff on foreign tobacco entering China. While this tariff has been reduced from 64% in 1999 to 10% in 2004, it still has led to local, Chinese cigarettes being preferred over foreign cigarettes because of their lower cost.

Every year 6.7 million tons of tobacco are produced throughout the world. The top producers of tobacco are China (39.6%), India (8.3%), Brazil (7.0%) and the United States (4.6%).

Cotton

Cotton is a soft, fluffy staple fibre that grows in a boll, or protective capsule, around the seeds of cotton plants of the genus *Gossypium*. The fibre is almost pure cellulose. Under natural conditions, the cotton bolls will tend to increase the dispersion of the seeds.

The plant is a shrub native to tropical and subtropical regions around the world, including the Americas, Africa, and India. The greatest diversity of wild cotton species is found in Mexico, followed by Australia and Africa. Cotton was independently domesticated in the Old and New Worlds. The English name derives from the Arabic *(al) qumn* ÞõØúä, which began to be used circa 1400 AD. The Spanish word, "algodón", is likewise derived from the Arabic. The fibre is most often spun into yarn or thread and used to make a soft, breathable textile. The use of cotton for fabric is known to date to prehistoric times; fragments of cotton fabric dated from 5000 BC have been excavated in Mexico and the Indus Valley Civilization (modern day Pakistan and some parts of India). Although cultivated since antiquity, it was the invention of the cotton gin that so lowered the cost of production that led to its widespread use, and it is the most widely used natural fibre cloth in clothing today.

Current estimates for world production are about 25 million tonnes or 110 million bales annually, accounting for 2.5% of the world's arable land. China is the world's largest producer of cotton, but most of this is used domestically. The United States has been the largest exporter for many years. In the United States, cotton is usually measured in bales, which measure approximately 0.48 cubic metres (17 cubic feet) and weigh 226.8 kilograms (500 pounds).

Types

There are four commercially grown species of cotton, all domesticated in antiquity:

- *Gossypium hirsutum* – upland cotton, native to Central America, Mexico, the Caribbean and southern Florida, (90% of world production)
- *Gossypium barbadense* – known as extra-long staple cotton, native to tropical South America (8% of world production)
- *Gossypium arboreum* – tree cotton, native to India and Pakistan (less than 2%)
- *Gossypium herbaceum* – Levant cotton, native to southern Africa and the Arabian Peninsula (less than 2%)

The two New World cotton species account for the vast majority of modern cotton production, but the two Old World species were widely used before the 1900s. While cotton fibres occur naturally in colours of white, brown, pink and green, fears of contaminating the genetics of white cotton have led many cotton-growing locations to ban the growing of coloured cotton varieties, which remain a speciality product.

History

Cotton was used in the Old World at least 7,000 years ago (5th millennium BC). Evidence of cotton use has been found at the site of Mehrgarh, where early cotton threads have been preserved in copper beads. Cotton cultivation became more widespread during the Indus Valley Civilization, which covered parts of modern eastern Pakistan and northwestern India. The Indus cotton industry was well developed and some methods used in cotton spinning and fabrication continued to be used until the industrialization of India. Between 2000 and 1000 BC cotton became widespread across much of India. For example, it has been found at the site of Hallus in Karnataka dating from around 1000 BC. Cotton fabrics discovered in a cave near Tehuacán, Mexico have been dated to around 5800 BC, although it is difficult to know for certain due to fibre decay. Other sources date the domestication of cotton in Mexico to approximately 5000 to 3000 BC.

Figure: *Cotton plants as imagined and drawn by John Mandeville in the 14th century*

The Greeks and the Arabs were not familiar with cotton until the Wars of Alexander the Great, as his contemporary Megasthenes told Seleucus I Nicator of "there being trees on which wool grows" in "Indica". This might actually be a reference to the 'tree cotton', Gossypium arboreum, which is a native of the Indian subcontinent.

According to the *Columbia Encyclopedia*:

Cotton has been spun, woven, and dyed since prehistoric times. It clothed the people of ancient India, Egypt, and China. Hundreds of years before the Christian era, cotton textiles were woven in India with matchless skill, and their use spread to the Mediterranean countries.

In Iran (Persia), the history of cotton dates back to the Achaemenid era (5th century BC); however, there are few sources about the planting of cotton in pre-Islamic Iran. The planting of cotton was common in Merv, Ray and Pars of Iran. In the poems of Persian poets, especially Ferdowsi's Shahname, there are references to cotton ("panbe" in Persian). Marco Polo (13th century) refers to the major products of Persia, including cotton. John Chardin, a French traveler of 17th century, who had visited the Safavid Persia, has approved the vast cotton farms of Persia.

During the Han dynasty, cotton was grown by non Chinese peoples in the southern Chinese province of Yunnan.

In Peru, cultivation of the indigenous cotton species *Gossypium barbadense* was the backbone of the development of coastal cultures, such as the Norte Chico, Moche and Nazca. Cotton was grown upriver, made into nets and traded with fishing villages along the coast for large supplies of fish. The Spanish who came to Mexico and Peru in the early 16th century found the people growing cotton and wearing clothing made of it.

During the late medieval period, cotton became known as an imported fibre in northern Europe, without any knowledge of how it was derived, other than that it was a plant. Because Herodotus had written in his *Histories*, Book III, 106, that in India trees grew in the wild producing wool, it was assumed that the plant was a tree, rather than a shrub. This aspect is retained in the name for cotton in several Germanic languages, such as German *Baumwolle*, which translates as "tree wool" (*Baum* means "tree"; *Wolle* means "wool"). Noting its similarities to wool, people in the region could only imagine that cotton must be produced by plant-borne sheep. John Mandeville, writing in 1350, stated as fact the now-preposterous belief: "There grew there [India] a wonderful tree which bore tiny lambs on the endes of its branches. These branches were so pliable that they bent down to allow the lambs to feed when they are hungrie [*sic*]." By the end of the 16th century, cotton was cultivated throughout the warmer regions in Asia and the Americas.

Figure: *The Vegetable Lamb of Tartary*

India's cotton-processing sector gradually declined during British expansion in India and the establishment of colonial rule during the late 18th and early 19th centuries. This was largely due to aggressive colonialist mercantile policies of the British East India Company, which made cotton processing and manufacturing workshops in India uncompetitive. Indian markets were increasingly forced to supply only raw cotton and were forced, by British-imposed law, to purchase manufactured textiles from Britain.

Industrial Revolution in Britain

The advent of the Industrial Revolution in Britain provided a great boost to cotton manufacture, as textiles emerged as Britain's leading export. In 1738, Lewis Paul and John Wyatt, of Birmingham, England, patented the roller spinning machine, and the flyer-and-bobbin system for drawing cotton to a more even thickness using two sets of rollers that travelled at different speeds. Later, the invention of the James Hargreaves' spinning jenny in 1764,Richard Arkwright's spinning frame in 1769 and Samuel Crompton's spinning mule in 1775 enabled British spinners to produce cotton yarn at much higher rates.

From the late 18th century onwards, the British city of Manchester acquired the nickname *"Cottonopolis"* due to the cotton industry's omnipresence within the city, and Manchester's role as the heart of the global cotton trade. Production capacity in Britain and the United States was improved by the invention of the cotton gin by the American Eli Whitney in 1793. Before the development of cotton gins, the cotton fibres had to be pulled from the seeds tediously by hand. By the late 1700s a number of crude ginning machines had been developed. However, to produce a bale of cotton required over 600 hours of human labour, making large scale production uneconomical in the United States, even with the use of humans as slave labour.

The gin that Whitney manufactured (the Holmes design) reduced the hours down to just a dozen or so per bale. Although Whitney patented his own design for a cotton gin, he manufactured a prior design from Henry Odgen Holmes, for which Holmes filed a patent in 1796. Improving technology and increasing control of world markets allowed British traders to develop a commercial chain in which raw cotton fibres were (at first) purchased from colonial plantations, processed into cotton cloth in the mills of Lancashire, and then exported on British ships to captive colonial markets in West Africa, India, and China (via Shanghai and Hong Kong).

By the 1840s, India was no longer capable of supplying the vast quantities of cotton fibres needed by mechanized British factories, while shipping bulky, low-price cotton from India to Britain was time-consuming and expensive. This, coupled with the emergence of American cotton as a superior type (due to the longer, stronger fibres of the two domesticated native American species, *Gossypium hirsutum* and *Gossypium barbadense*), encouraged British traders to purchase cotton from plantations in the United States and the Caribbean.

By the mid-19th century, "King Cotton" had become the backbone of the southern American economy. In the United States, cultivating and harvesting cotton became the leading occupation of slaves.

During the American Civil War, American cotton exports slumped due to a Union blockade on Southern ports, and also because of a strategic decision by the Confederate government to cut exports, hoping to force Britain to recognize the Confederacy or enter the war. This prompted the main purchasers of cotton, Britain and France, to turn to Egyptian cotton. British and French traders invested heavily in cotton plantations and the Egyptian government of Viceroy Isma'il took out substantial loans from European bankers and stock exchanges.

After the American Civil War ended in 1865, British and French traders abandoned Egyptian cotton and returned to cheap American exports, sending Egypt into a deficit spiral that led to the country declaring bankruptcy in 1876, a key factor behind Egypt's occupation by the British Empire in 1882.

During this time, cotton cultivation in the British Empire, especially India, greatly increased to replace the lost production of the American South. Through tariffs and other restrictions, the British government discouraged the production of cotton cloth in India; rather, the raw fibre was sent to England for processing. The Indian Mahatma Gandhi described the process:

1. English people buy Indian cotton in the field, picked by Indian labour at seven cents a day, through an optional monopoly.
2. This cotton is shipped on British ships, a three-week journey across the Indian Ocean, down the Red Sea, across the Mediterranean, through Gibraltar, across the Bay of Biscay and the Atlantic Ocean to London. One hundred per cent profit on this freight is regarded as small.
3. The cotton is turned into cloth in Lancashire. You pay shilling wages instead of Indian pennies to your workers. The English worker not only has the advantage of better wages, but the steel companies of England get the profit of building the factories and machines. Wages; profits; all these are spent in England.

4. The finished product is sent back to India at European shipping rates, once again on British ships. The captains, officers, sailors of these ships, whose wages must be paid, are English. The only Indians who profit are a few lascars who do the dirty work on the boats for a few cents a day.
5. The cloth is finally sold back to the kings and landlords of India who got the money to buy this expensive cloth out of the poor peasants of India who worked at seven cents a day.

In the United States, Southern cotton provided capital for the continuing development of the North. The cotton produced by enslaved African Americans not only helped the South, but also enriched Northern merchants. Much of the Southern cotton was transshipped through the northern ports.

Cotton remained a key crop in the Southern economy after emancipation and the end of the Civil War in 1865. Across the South, sharecropping evolved, in which free black farmers and landless white farmers worked on white-owned cotton plantations of the wealthy in return for a share of the profits. Cotton plantations required vast labour forces to hand-pick cotton, and it was not until the 1950s that reliable harvesting machinery was introduced into the South (prior to this, cotton-harvesting machinery had been too clumsy to pick cotton without shredding the fibres).

During the early 20th century, employment in the cotton industry fell, as machines began to replace labourers, and the South's rural labour force dwindled during the First and Second World Wars. Today, cotton remains a major export of the southern United States, and a majority of the world's annual cotton crop is of the long-staple American variety.

Cultivation

Successful cultivation of cotton requires a long frost-free period, plenty of sunshine, and a moderate rainfall, usually from 600 to 1,200 mm (24 to 47 in). Soils usually need to be fairly heavy, although the level of nutrients does not need to be exceptional. In general, these conditions are met within the seasonally dry tropics and subtropics in the Northern and Southern hemispheres, but a large proportion of the cotton grown today is cultivated in areas with less rainfall that obtain the water from irrigation. Production of the crop for a given year usually starts soon after harvesting the preceding autumn. Planting time in spring in the Northern hemisphere varies from the beginning of February to the beginning of June.

The area of the United States known as the South Plains is the largest contiguous cotton-growing region in the world. While dryland (non-irrigated) cotton is successfully grown in this region, consistent yields are only produced with heavy reliance on irrigation water drawn from the Ogallala Aquifer. Since cotton is somewhat salt and drought tolerant, this makes it an attractive crop for arid and semiarid regions. As water resources get tighter around the world, economies that rely on it face difficulties and conflict, as well as potential environmental problems. For example, improper cropping and irrigation practices have led to desertification in areas of Uzbekistan, where cotton is a major export. In the days of the Soviet Union, the Aral Sea was tapped for agricultural irrigation, largely of cotton, and now salination is widespread.

Cotton can also be cultivated to have colours other than the yellowish off-white typical of modern commercial cotton fibres. Naturally coloured cotton can come in red, green, and several shades of brown.

Genetic Modification

Genetically modified (GM) cotton was developed to reduce the heavy reliance on pesticides. The bacterium *Bacillus thuringiensis* (Bt) naturally produces a chemical harmful only to a small fraction of insects, most notably the larvae of moths and butterflies, beetles, and flies, and harmless to other forms of life. The gene coding for Bt toxin has been inserted into cotton, causing cotton, called Bt cotton, to produce this natural insecticide in its tissues. In many regions, the main pests in commercial cotton are lepidopteran larvae, which are killed by the Bt protein in the transgenic cotton they eat. This eliminates the need to use large amounts of broad-spectrum insecticides to kill lepidopteran pests (some of which have developed pyrethroid resistance). This spares natural insect predators in the farm ecology and further contributes to noninsecticide pest management.

But cotton is ineffective against many cotton pests, however, such as plant bugs, stink bugs, and aphids; depending on circumstances it may still be desirable to use insecticides against these. A 2006 study done by Cornell researchers, the Centre for Chinese Agricultural Policy and the Chinese Academy of Science on Bt cotton farming in China found that after seven years these secondary pests that were normally controlled by pesticide had increased, necessitating the use of pesticides at similar levels to non-Bt cotton and causing less profit for farmers because of the extra expense of GM seeds. However, a 2009 study by the Chinese Academy of Sciences, Stanford University and Rutgers University refuted this. They concluded that the GM cotton effectively controlled bollworm. The secondary pests were mostly miridae (plant bugs) whose increase was related to local temperature and rainfall and only continued to increase in half the villages studied. Moreover, the increase in insecticide use for the control of these secondary insects was far smaller than the reduction in total insecticide use due to Bt cotton adoption. A 2012 Chinese study concluded that Bt cotton halved the use of pesticides and doubled the level of ladybirds, lacewings and spiders. The International Service for the Acquisition of Agri-biotech Applications (ISAAA) said that, worldwide, GM cotton was planted on an area of 25 million hectares in 2011. This was 69% of the worldwide total area planted in cotton.

GM cotton acreage in India grew at a rapid rate, increasing from 50,000 hectares in 2002 to 10.6 million hectares in 2011. The total cotton area in India was 12.1 million hectares in 2011, so GM cotton was grown on 88% of the cotton area. This made India the country with the largest area of GM cotton in the world. A long-term study on the economic impacts of Bt cotton in India, published in the Journal PNAS in 2012, showed that Bt cotton has increased yields, profits, and living standards of smallholder farmers.

The U.S. GM cotton crop was 4.0 million hectares in 2011 the second largest area in the world, the Chinese GM cotton crop was third largest by area with 3.9 million hectares and Pakistan had the fourth largest GM cotton crop area of 2.6 million hectares in 2011. The initial introduction of GM cotton proved to be a success in Australia – the yields were equivalent to the non transgenic varieties and the crop used much less pesticide to produce (85% reduction). The subsequent introduction of a second variety of GM cotton led to increases in GM cotton production until 95% of the Australian cotton crop was GM in 2009 making Australia the country with the fifth largest GM cotton crop in the world. Other GM cotton growing countries in 2011 were Argentina, Myanmar, Burkina Faso, Brazil, Mexico, Colombia, South Africa and Costa Rica.

Cotton has been genetically modified for resistance to glyphosate a broad-spectrum herbicide discovered by Monsanto which also sells some of the Bt cotton seeds to farmers. There are also a number of other cotton seed companies selling GM cotton around the world. About 62% of the GM cotton grown from 1996 to 2011 was insect resistant, 24% stacked product and 14% herbicide resistant.

Cotton has gossypol, a toxin that makes it inedible. However, scientists have silenced the gene that produces the toxin, making it a potential food crop.

Organic Production

Organic cotton is generally understood as cotton, from plants not genetically modified, that is certified to be grown without the use of any synthetic agricultural chemicals, such as fertilizers or pesticides. Its production also promotes and enhances biodiversity and biological cycles. United States cotton plantations are required to enforce the National Organic Programme (NOP). This institution determines the allowed practices for pest control, growing, fertilizing, and handling of organic crops. As of 2007, 265,517 bales of organic cotton were produced in 24 countries, and worldwide production was growing at a rate of more than 50% per year.

Pests and Weeds

The cotton industry relies heavily on chemicals, such as herbicides, fertilizers and insecticides, although a very small number of farmers are moving toward an organic model of production, and organic cotton products are now available for purchase at limited locations. These are popular for baby clothes and diapers. Under most definitions, organic products do not use genetic engineering. All natural cotton products are known to be both sustainable and hypoallergenic.

Historically, in North America, one of the most economically destructive pests in cotton production has been the boll weevil. Due to the US Department of Agriculture's highly successful Boll Weevil Eradication Programme (BWEP), this pest has been eliminated from cotton in most of the United States. This programme, along with the introduction of genetically engineered Bt cotton (which contains a bacterial gene that codes for a plant-produced protein that is toxic to a number of pests such as cotton bollworm and pink bollworm), has allowed a reduction in the use of synthetic insecticides.

Other significant global pests of cotton include the pink bollworm, *Pectinophora gossypiella*; the chili thrips, *Scirtothrips dorsalis*; the cotton seed bug, *Oxycarenus hyalinipennis*; the tarnish plant bug, *Lygus lineolaris*; and the fall armyworm, *Spodoptera frugiperda, Xanthomonas citri subsp. malvacearum*.

Harvesting

Most cotton in the United States, Europe, and Australia is harvested mechanically, either by a cotton picker, a machine that removes the cotton from the boll without damaging the cotton plant, or by a cotton stripper, which strips the entire boll off the plant. Cotton strippers are used in regions where it is too windy to grow picker varieties of cotton, and usually after application of a chemical defoliant or the natural defoliation that occurs after a freeze. Cotton is a perennial crop in the tropics, and without defoliation or freezing, the plant will continue to grow.

Cotton continues to be picked by hand in developing countries.

Uses

Cotton is used to make a number of textile products. These include terrycloth for highly absorbent bath towels and robes; denim for blue jeans; cambric, popularly used in the manufacture of blue work shirts (from which we get the term "blue-collar"); and corduroy, seersucker, and cotton twill. Socks, underwear, and most T-shirts are made from cotton. Bed sheets often are made from cotton. Cotton also is used to make yarn used in crochet and knitting. Fabric also can be made from recycled or recovered cotton that otherwise would be thrown away during the spinning, weaving, or cutting process. While many fabrics are made completely of cotton, some materials blend cotton with other fibres, including rayon and synthetic fibres such as polyester. It can either be used in knitted or woven fabrics, as it can be blended with elastine to make a stretchier thread for knitted fabrics, and apparel such as stretch jeans.

In addition to the textile industry, cotton is used in fishing nets, coffee filters, tents, explosives manufacture, cotton paper, and in bookbinding. The first Chinese paper was made of cotton fibre. Fire hoses were once made of cotton.

The cottonseed which remains after the cotton is ginned is used to produce cottonseed oil, which, after refining, can be consumed by humans like any other vegetable oil. The cottonseed meal that is left generally is fed to ruminant livestock; the gossypol remaining in the meal is toxic to monogastric animals. Cottonseed hulls can be added to dairy cattle rations for roughage. During the American slavery period, cotton root bark was used in folk remedies as an abortifacient, that is, to induce a miscarriage. Gossypol was one of the many substances found in all parts of the cotton plant and it was described by the scientists as 'poisonous pigment'. It also appears to inhibit the development of sperm or even restrict the mobility of the sperm. Also, it is thought to interfere with the menstrual cycle by restricting the release of certain hormones.

Cotton linters are fine, silky fibres which adhere to the seeds of the cotton plant after ginning. These curly fibres typically are less than D $_{8}$ inch (3.2 mm) long. The term also may apply to the longer textile fibre staple lint as well as the shorter fuzzy fibres from some upland species. Linters are traditionally used in the manufacture of paper and as a raw material in the manufacture of cellulose. In the UK, linters are referred to as "cotton wool". This can also be a refined product (*absorbent cotton* in U.S. usage) which has medical, cosmetic and many other practical uses. The first medical use of cotton wool was by Dr. Joseph Sampson Gamgee at the Queen's Hospital (later the General Hospital) in Birmingham, England.

Shiny cotton is a processed version of the fibre that can be made into cloth resembling satin for shirts and suits. However, it is hydrophobic (does not absorb water easily), which makes it unfit for use in bath and dish towels (although examples of these made from shiny cotton are seen).

The name Egyptian cotton is broadly associated with quality products, however only a small percentage of Egyptian cotton production is actually of superior quality. Most products bearing the name are not made with the finest cottons from Egypt.

Pima cotton is often compared to Egyptian cotton, as both are used in high quality bed sheets and other cotton products. It is even considered superior by some authorities. Pima cotton is grown in the American southwest.

International Trade

The largest producers of cotton, currently (2009), are China and India, with annual production of about 34 million bales and 27 million bales, respectively; most of this production is consumed by their respective textile industries. The largest exporters of raw cotton are the United States, with sales of $4.9 billion, and Africa, with sales of $2.1 billion. The total international trade is estimated to be $12 billion. Africa's share of the cotton trade has doubled since 1980. Neither area has a significant domestic textile industry, textile manufacturing having moved to developing nations in Eastern and South Asia such as India and China. In Africa, cotton is grown by numerous small holders. Dunavant Enterprises, based in Memphis, Tennessee, is the leading cotton broker in Africa, with hundreds of purchasing agents. It operates cotton gins in Uganda, Mozambique, and Zambia. In Zambia, it often offers loans for seed and expenses to the 180,000 small farmers who grow cotton for it, as well as advice on farming methods. Cargill also purchases cotton in Africa for export.

The 25,000 cotton growers in the United States of America are heavily subsidized at the rate of $2 billion per year although China now provides the highest overall level of cotton sector support. The future of these subsidies is uncertain and has led to anticipatory expansion of cotton brokers' operations in Africa. Dunavant expanded in Africa by buying out local operations. This is only possible in former British colonies and Mozambique; former French colonies continue to maintain tight monopolies, inherited from their former colonialist masters, on cotton purchases at low fixed prices.

Vineyard

A vineyard is a plantation of grape-bearing vines, grown mainly for winemaking, but also raisins, table grapes and non-alcoholic grape juice. The science, practice and study of vineyard production is known as viticulture. A vineyard is often characterised by its terroir, a French term loosely translating as "a sense of place" that refers to the specific geographical and geological characteristics of grapevine plantations, which may be imparted in the wine.

The earliest evidence of wine production dates from between 6000 and 5000 BC. Wine making technology improved considerably with the ancient Greeks but it wasn't until the end of the Roman Empire that cultivation techniques as we know them were common throughout Europe. In medieval Europe the Church was a staunch supporter of wine, which was necessary for the celebration of the Mass. During the lengthy instability of the Middle Ages, the monasteries maintained and developed viticultural practices, having the resources, security, stability and interest in improving the quality of their vines. They owned and tended the best vineyards in Europe and *vinum theologium* was considered superior to all others.

European vineyards were planted with a wide variety of the *Vitis vinifera* grape. However, in the late 19th century, the entire species was nearly destroyed by the plant louse *phylloxera* accidentally introduced to Europe from North America. Native American grapevines include varieties such as *Vitis labrusca*, which is resistant to the bug. *Vitis vinifera* varieties were saved by being grafted onto the rootstock of native American varieties, although there is still no remedy for *phylloxera*, which remains a danger to any vineyard not planted with grafted rootstock.

Modern Practices

The quest for vineyard efficiency has produced a bewildering range of systems and techniques in recent years. Due to the often much more fertile New World growing conditions, attention has focussed heavily on managing the vine's more vigorous growth. Innovation in *palissage* (training of the vine, usually along a trellis, and often referred to as "canopy management") and pruning and thinning methods (which aim to optimize the Leaf Area/Fruit (LA/F) ratio relative to a vineyard's microclimate) have largely replaced more general, traditional concepts like "yield per unit area" in favour of "maximizing yield of desired quality". Many of these new techniques have since been adopted in place of traditional practice in the more progressive of the so-called "Old World" vineyards.

Other recent practices include spraying water on vines to protect them from sub-zero temperatures (aspersion), new grafting techniques, soil slotting, and mechanical harvesting. Such techniques have made possible the development of wine industries in New World countries such as Canada. Today there is increasing interest in developing organic, ecologically sensitive and sustainable vineyards. Biodynamics has become increasingly popular in viticulture. The use of drip irrigation in recent years has expanded vineyards into areas which were previously unplantable.

For well over half a century Cornell University, the University of California, Davis, and California State University, Fresno, among others, have been conducting scientific experiments to improve viticulture and educating practitioners. The research includes developing improved grape varieties and investigating pest control. The International Grape Genome Programme is a multi-national effort to discover a genetic means to improving quality, increasing yield and providing a "natural" resistance to pests.

The implementation of mechanical harvesting is often stimulated by changes in labour laws, labour shortages, and bureaucratic complications. It can be expensive to hire labour for short periods of time, which does not square well with the need to reduce production costs and harvest quickly, often at night. However, very small vineyards, incompatible widths between rows of grape vines and steep terrain hinder the employment of machine harvesting even more than the resistance of traditional views which reject such harvesting.

Current Trends

Numbers of New World vineyard plantings have been increasing almost as fast as European vineyards are being uprooted. Between 1990 and 2003, U.S. vineyards increased from 292,000 acres (1,180 km^2) to 954,000 acres (3,860 km^2), while Australian vineyard numbers more than doubled from 146,000 acres (590 km^2) to 356,000 acres (1,440 km^2) and Chilean vineyards grew from 161,500 acres (654 km^2) to 415,000 acres (1,680 km^2). The size of individual vineyards in the New World is significant. Europe's 1.6 million vineyards are an average of 0.2 square kilometres each, while the average Australian vineyard is 0.5

square kilometres, providing considerable economies of scale. Exports to Europe from New World growers increased by 54% in the six years up to 2006.

There are also changes in the kinds of grapes grown. For example, in Chile, large areas of low-quality grapes have been replaced with such grapes as Chardonnay and Cabernet Sauvignon. Grape changes are often in response to changing consumer demand but sometimes result from vine pull schemes designed to promote vineyard change. Alternatively, the development of "T" budding now permits the grafting of a different grape variety onto existing rootstock in the vineyard, making it possible to switch varieties within a two-year period.

Local legislation often dictates which varieties are selected, how they are grown, whether vineyards can be irrigated and exactly when grapes can be harvested, all of which in serves to reinforce tradition. Of course, changes in the law can change which grapes are planted. For example, during Prohibition in the U.S. (1920–1933), vineyards in California expanded sevenfold to meet the increasing demand for home-brewing. However, they were largely planted in varieties with tough skins that could be transported across the country to home wine-makers and the resulting wine was of low quality.

Terroir

Figure: *Vines growing in volcanic lapilli in the La Geria region of Lanzarote. The low, curved walls are traditionally used to protect the vines from the constant wind.*

Terroir refers to the combination of natural factors associated with any particular vineyard. These factors include such things as soil, underlying rock, altitude, slope of hill or terrain, orientation toward the sun, and microclimate (typical rain, winds, humidity, temperature variations, etc.) No two vineyards have exactly the same terroir, although any difference in the resulting wine may be virtually undetectable.

Vineyards are often on hillsides and on soil of marginal value to other plants. A common saying is that "the worse the soil, the better the wine." Planting on hillsides, especially those facing north (in the southern hemisphere) or south (in the northern hemisphere), is most often in an attempt to maximize the amount of sunlight that falls on the vineyard. For this reason some of the best wines come from vineyards planted on quite steep hills, conditions which would make most other agricultural products uneconomic. The stereotypical vineyard site for wine grapes (in the Northern hemisphere) is a hillside in a dry climate with a southern exposure, good drainage to reduce unnecessary water uptake, and balanced pruning to force the vine to put more of its energy into the fruit, rather than foliage.

The terroir philosophy is predominately French, the flavour and character of the place defining the individuality and the special attributes of wines and combined with hundreds of years of the finest wine making traditions, terroir gives wines their distinctive taste and signature.

Vignette

A vignette is a 500 square metre vineyard which is part of a larger consolidated vineyard. Investors purchase a piece of land within a vineyard, and outsource the grape maintenance and production operations to an outside grape grower or wine producers. Because they are contracting under a co-operative structure, they benefit from economies of scale and hence cheaper labour and operational costs.

Viticulture

Viticulture (from the Latin word for *vine*) is the science, production, and study of grapes. It deals with the series of events that occur in the vineyard. When the grapes are used for winemaking, it is also known as viniculture. It is a branch of the science of horticulture. While the native territory of *Vitis vinifera*, the common grape vine, ranges from Western Europe to the Persian shores of the Caspian Sea, the vine has demonstrated high levels of adaptability and will sometimes mutate to accommodate a new environment after its introduction. For this reason, viticulture can be found on every continent except Antarctica.

Duties of the viticulturist include monitoring and controlling pests and diseases, fertilizing, irrigation, canopy management, monitoring fruit development and characteristics, deciding when to harvest, and vine pruning during the winter months. Viticulturists are often intimately involved with winemakers, because vineyard management and the resulting grape characteristics provide the basis from which winemaking can begin. A great number of varieties are now approved in the European Union as true grapes for wine-growing and viticulture, because of better characteristics for diseases.

The Grape Vine

The vast majority of the world's wine producing regions are found between the temperate latitudes of 30° and 50° in each hemisphere. Within these bands, the annual mean temperatures are between 10 and 20 °C (50 and 68 °F). The presence of large bodies of water and mountain ranges can have positive effects on the climate and vines. Nearby lakes and rivers can serve as protection for drastic temperature drops at night by releasing the heat that the water has stored during the day to warm the vines.

The Berry

The grape is classified as a berry. On the vine, grapes are organized through systems known as clusters. Grape clusters can vary in compactness which can result in long clusters (resulting in the grapes spreading out) or short clusters (resulting in grapes packed together). In some grape species, clusters ripen collectively which allow them to be harvested together. For others, grapes may ripen individually within a cluster. Each grape berry contains a pedicel which attaches to the rachis. The main function of the rachis is to allow the grapes receive their water and nutrients. The pollination and fertilization of grapes results in one to four seeds within the berry. When fertilization does not occur, seedless grapes are formed, which are sought after for the production of raisins. Regardless of pollination and fertilization, most plants will produce around 100 to 200 grapes.

The skin of the grape accounts for 5 to 20% of the total weight of a grape depending on the variety. When grape skin ripens, it contains the majority of the aromatic substances and tannin. These factors become important in winemaking for methods including colour extraction or aroma dissolution. Although the skin contains the majority of the tannin, small percentages can be found throughout the grape and during all of its developmental stages. However, the tannin's most important role is during the grape's ripening stage as its function is to formulate colour and body shape.

Growing Vines

Although many factors can affect the overall quality of a grape vine, the three most important are climate, slope, and soil.

Climate

Climate is the most significant factor in determining a grape's inherent qualities. Each grape species has a uniquely preferred environment for ideal growing. Because climates vary from country to country, selecting the best strain is an important decision in grape cultivation. In addition, because climatic factors such as temperature and rain can be unpredictable and uncontrollable, each year will produce unique qualities and yields of grapes.

Grape vines need approximately 1300–1500 hours of sunshine during the growing season and around 690 millimetres (27 in) of rainfall throughout the year in order to produce grapes suitable for winemaking. In ideal circumstances, the vine will receive most of the rainfall during the winter and spring months: rain at harvesttime can create many hazards, such as fungal diseases and berry splitting. The optimum weather during the growing season is a long, warm summer that allows the grapes the opportunity to ripen fully and to develop a balance between the levels of acids and sugars in the grape. Hot and sunny climates have a frost-free growing season of 200 days or more. These climates allow grapes to ripen faster with higher sugar levels and lower acidity. Cooler climates have a frost-free growing season of around 150–160 days. Cooler seasons force the grapes to ripen earlier which produces a fresher and more acidic harvest. In general, the average yearly temperature for most crops should average around 15 °C (59 °F) in order for the highest quality to be achieved in each grape.

Summer: Ideal temperatures in summer average around 22 °C (72 °F). Ideal summer temperatures enable fruits to ripen. Temperature and sunshine are the most important factors in ripening.

Winter: Ideal temperatures in winter average around 3 °C (37 °F). Ideal winter temperatures are necessary to allow grape vines to enter their resting phase. If temperatures fall too low, the crops can be injured.

Spring and Fall: Spring and fall are critical seasons for grape development, because the plants are susceptible to frost damage, which can injure the fruiting buds. Wet weather in spring can increase the odds of mildew formation. To prevent mildew, some farms introduce devices such as heaters or large fans in vineyards. However, such solutions can be costly.

Slope

Hillsides and slopes are preferred over flatter terrain: vines growing on a slope can receive a greater intensity of the sun's rays, with sunshine falling on an angle perpendicular to the hillside. In flatter terrain, the intensity of the sunlight is diluted as it spreads out across a wider surface area. Small slopes that are elevated above surrounding ground are the best and safest places for crops, because these small elevations are less prone to frost. Additionally, a slope affords better drainage, obviating the possibility that the vine might sit in overly moist soil. In cooler regions of the northern hemisphere, south-facing slopes receive more hours of sunlight and are preferred; in warmer climes, north-facing slopes are preferred. In the southern hemisphere, these orientations are reversed.

Soil

Quality soil is important to allow plants to have better root systems. The growth and health of a vine can be affected if the soil quality is poor. Different grape species prefer various soil conditions, although there are general quality factors. Favourable soil conditions include: aeration, loose texture, good drainage, and moderate fertility. Drainage factors are cited as the most important soil characteristic to affect grape vine growth. When root growth is restricted due to bad soil, vine growth and fruit yields lessen and plant survival rates can dip to only a few years.

Hazards

There are many hazards that a viticulturist needs to be aware of when growing vines. These hazards can have an adverse effect on the wine produced from the grape or kill the vine itself. When the vine is flowering it is very susceptible to weather hazards such as strong winds and hail. Cold temperatures during this period can also bring the onset of *millerandage* which produces clusters with no seeds and varying sizes. Too much heat can have the opposite reaction and produce Coulure that causes grape clusters to either drop to the ground or not fully develop.

Viticultural hazards include:

- Oidium: A powdery mildew which can attack all green parts of the vine. If left untreated, odium can be terminal for the plant. It thrives in cooler temperatures and in the shade. Some North American vine species have evolved to show resistance to the mildew (3:37).
- Peronospera: Also known as Downy Mildew, thrives in high temperatures and humidity and produces stains on leaves. Peronospera can be treated by spraying plants with copper sulphate. Most American vines are resistant, excluding Vitis vinifera.

- Fanleaf: can lead to deformity, yellowing of the leaves, and smaller crop yields. Fanleaf is a virus spread by nematodes which breed in the vine stem. There is no cure for the plant, and the best action to take is to remove the infected plants and allow remaining roots in the soil to rot.
- Powdery mildew
- Frost
- Oidium
- Phylloxera
- Plant virus

Green Harvest

A green harvest is the removal of immature grape bunches, typically for the purpose of decreasing yield. In French, it is known as a *vendange verte*.

Green harvesting is a relatively modern practice most often used to produce fine wine. Removing the tiny, immature grapes while they are still green induces the vine to put all its energy into developing the remaining grapes. In theory this results in better ripening and the development of more numerous and mature flavour compounds. In the absence of a green harvest, a healthy, vigorous vine can produce dilute, unripe grapes.

Many traditionally renowned regions have natural conditions that suppress excess vigor. Examples include the gravelly soil of Bordeaux, the often cool climate of Burgundy, and the meager rainfall of Rioja. In these regions, the vine is prevented from producing too many grapes without human intervention. However, in regions with fertile soil, copious sunlight, and irrigation, the vine can generate huge quantities of characterless grapes. One solution is a green harvest. After fruit set, the quantity of grapes that will result from a vineyard can be estimated. Often the grower has a target yield in mind, measured in tons per acre or hectoliters per hectare. A portion of the grape bunches are cut off, to leave approximately the correct amount.

In Europe, many appellations restrict the yield permitted from a given area, so there is even more incentive to perform green harvesting when presented with excess crop. Often the excess must be sold for a pittance and used for industrial alcohol production rather than wine.

While the concept of thinning or sacrificing part of the grape crop, i.e. green harvesting, with the aim of improving the quality of the remaining grapes, predates modern critics, the practice has increased in recent times in vineyards found in California and areas where the grapes grow easily. (McCoy)

Field Blend

A field blend is a wine that is produced from two or more different grape varieties interplanted in the same vineyard. In the days before precise varietal identification, let alone rigorous clonal selection, a vineyard might be planted by taking cuttings from another vineyard and therefore approximately copying its genetic makeup. This meant that one vine could be Zinfandel and the next Carignan. When making wine with little equipment to spare for separate vinification of different varieties, field blends allowed effortless, though inflexible, blending.

Fermentation tanks are now cheap enough that the field blend is an anachronism, and almost all wines are assembled by blending from smaller, individual lots. However, in California some of the oldest (and lowest-yielding) Zinfandel comes from vineyards that are field-blended. Ridge Vineyards owns the Lytton Springs vineyards in Sonoma County, which were planted from 1900 to 1905 in what Ridge calls "a traditional field blend of about seventy percent Zinfandel, twenty percent Petite Sirah, and ten percent Grenache and Carignan."

Gemischter Satz (*Mixed set*) is a wine term in German equivalent to a field blend, which means that grapes of different varieties are planted, harvested and vinified together. In older times, this was common, but the practice has almost stopped. It is, however, a speciality of Vienna.

Pineapple

The pineapple (*Ananas comosus*) is a tropical plant with edible multiple fruit consisting of coalesced berries, and the most economically significant plant in the Bromeliaceae family. Pineapples may be cultivated from a crown cutting of the fruit, possibly flowering in 20–24 months and fruiting in the following six months. Pineapple does not ripen significantly post-harvest.

Pineapples are consumed fresh, cooked, juiced, and preserved, and are found in a wide array of cuisines. In addition to consumption, in the Philippines the pineapple's leaves are used to produce the textile fibre piña- employed as a component of wall paper and furnishings, amongst other uses.

Botany

The pineapple is a herbaceous perennial which grows to 1.0 to 1.5 metres (3.3 to 4.9 ft) tall, although sometimes it can be taller. In appearance, the plant itself has a short, stocky stem with tough, waxy leaves. When creating its fruit, it usually produces up to 200 flowers, although some large-fruited cultivars can exceed this. Once it flowers, the individual fruits of the flowers join together to create what is commonly referred to as a pineapple. After the first fruit is produced, side shoots (called 'suckers' by commercial growers) are produced in the leaf axils of the main stem.

These may be removed for propagation, or left to produce additional fruits on the original plant. Commercially, suckers that appear around the base are cultivated. It has 30 or more long, narrow, fleshy, trough-shaped leaves with sharp spines along the margins that are 30 to 100 centimetres (1.0 to 3.3 ft) long, surrounding a thick stem. In the first year of growth, the axis lengthens and thickens, bearing numerous leaves in close spirals. After 12 to 20 months, the stem grows into a spike-like inflorescence up to 15 cm (6 in) long with over 100 spirally arranged, trimerous flowers, each subtended by a bract. Flower colours vary, depending on variety, from lavender, through light purple to red.

Figure: *A pineapple flower in Iriomote, Japan*

The ovaries develop into berries which coalesce into a large, compact, multiple accessory fruit. The fruit of a pineapple is arranged in two interlocking helices, eight in one direction, thirteen in the other, each being a Fibonacci number.

Pineapple carries out CAM photosynthesis, fixing carbon dioxide at night and storing it as the acid malate and then releasing it during the day, aiding photosynthesis.

Pollination

Pollination is required for seed formation, but the presence of seeds negatively affects the quality of the fruit. In Hawaii, where pineapple is cultivated on an agricultural scale, importation of hummingbirds is prohibited for this reason. Certain bat-pollinated wild pineapples open their flowers only at night.

Culinary Uses

The flesh and juice of the pineapple are used in cuisines around the world. In many tropical countries, pineapple is prepared, and sold on roadsides as a snack. It is sold whole, or in halves with a stick inserted. Whole, cored slices with a cherry in the middle are a common garnish on hams in the West. Chunks of pineapple are used in desserts such as fruit salad, as well as in some savory dishes, including pizza toppings and a grilled ring on a hamburger. Crushed pineapple is used in yogurt, jam, sweets, and ice cream. The juice of the pineapple is served as a beverage, and is also as a main ingredient in such cocktails as the Piña colada.

Nutrition

Raw pineapple is an excellent source of manganese (76% Daily Value (DV) in a one US cup serving) and vitamin C (131% DV per cup serving). Mainly from its stem, pineapple contains a proteolytic enzyme, bromelain, which breaks down protein. If having sufficient bromelain content, raw pineapple juice may be used as a meat marinade and tenderizer. Pineapple enzymes can interfere with the preparation of some foods, such as jelly or other gelatin-based desserts, but would be destroyed during cooking and canning. The quantity of bromelain in the fruit is probably not significant, being mostly in the inedible stalk. Furthermore, an ingested enzyme like bromelain is unlikely to survive intact the proteolytic processes of digestion.

Cultivation

Southeast Asia dominates world production: in 2009, the Philippines produced 2.198 million tons and Thailand 1.894 million tons, while in the Americas, Brazil produced 1.471 million tons. Total world production in 2009 was 14.220 million tons. The primary exporters of fresh pineapples in 2001 were Costa Rica, 322,000 tons; Côte d'Ivoire, 188,000 tons; and the Philippines, 135,000 tons. Since 2000, the most common fresh pineapple fruit found in U.S. and European supermarkets is a low-acid hybrid that was developed in Hawaii in the early 1970s.

In commercial farming, flowering can be induced artificially, and the early harvesting of the main fruit can encourage the development of a second crop of smaller fruits. Once removed during cleaning, the top of the pineapple can be planted in soil and a new plant will grow. Slips and suckers are planted commercially.

Ethical and Environmental Concerns

Three-quarters of pineapples sold in Europe are grown in Costa Rica, where pineapple production is highly industrialised. Growers typically use 20 kg of pesticides per hectare in each growing cycle, a process that may affect soil quality and biodiversity. The pesticides – organophosphates, organochlorines and hormone disruptors – have the potential to affect workers' health and can contaminate local drinking water supplies. Many of these chemicals have potential to be carcinogens, and may be related to birth defects. Because of commercial pressures, many pineapple workers – 60% of whom are Nicaraguan – in Costa Rica are paid low wages. European supermarkets' price-reduction policies have lowered growers' incomes. One major pineapple producer contests these claims.

Cultivars

There are many cultivars. The leaves of the commonly grown 'Smooth Cayenne' are smooth and it is the most commonly grown worldwide. Many cultivars have become distributed from its origins in Paraguay and the southern part of Brazil, and later improved stocks were introduced into the Americas, the Azores, Africa, India, Malaysia and Australia. Varieties include:

- 'Hilo': A compact 1–1.5 kg (2–3 lb) Hawaiian variant of 'Smooth Cayenne', the fruit is more cylindrical and produces many suckers, but no slips.
- 'Kona Sugarloaf': 2.5–3 kg (5–6 lb), white flesh with no woodiness in the centre, cylindrical in shape, it has a high sugar content but no acid, an unusually sweet fruit.
- 'Natal Queen': 1–1.5 kg (2–3 lb), golden yellow flesh, crisp texture and delicate mild flavour, well-adapted to fresh consumption, keeps well after ripening, spiny leaves, grown in Australia, Malaysia, and South Africa
- 'Pernambuco' ('Eleuthera'): 1–2 kg (2–4 lb) with pale yellow to white flesh, sweet, melting and excellent for eating fresh, poorly adapted for shipping, spiny leaves, grown in Latin America
- 'Red Spanish': 1–2 kg (2–4 lb), pale yellow flesh with pleasant aroma, squarish in shape, well-adapted for shipping as fresh fruit to distant markets, spiny leaves, grown in Latin America
- 'Smooth Cayenne': 2.5–3 kg (5–6 lb), pale yellow to yellow flesh,

cylindrical in shape, high sugar and acid content, well-adapted to canning and processing, leaves without spines, this is the variety from Hawaii, and the most easily obtainable in US grocery stores. Both 73–114 and 73-50 are of this cultivar.

- Some Ananas species are grown as ornamentals for colour, novel fruit size and other esthetic qualities.

Traditional Medicine and Preliminary Research

Both the root and fruit may be eaten or applied topically as an anti-inflammatory or as a proteolytic agent. In some practices, it may be used to induce abortion or menstruation or as an antihelminthic agent. Bromelain purified from pineapple stem or fresh juice, then provided in the diet over six months, decreased the severity of colonic inflammation in mice with experimental colitis. Bromelain from pineapple has some potential against cancer mechanisms, as laboratory research showed that it causes autophagy in mammary carcinoma cells, stimulating turnover of MCF-7 cells through apoptosis.

Pests and Diseases

Pineapples are subject to a variety of diseases, the most serious of which is wilt disease vectored by mealybugs typically found on the surface of pineapples, but possibly in the closed blossom cups. Other diseases include pink disease, bacterial heart rot, anthracnose, fungal heart rot, root rot, black rot, butt rot, fruitlet core rot, and yellow spot virus. Pink disease is characterized by the fruit developing a brownish to black discolouration when heated during the canning process. The causal agents of pink disease are the bacteria *Acetobacter aceti*, *Gluconobacter oxydans*, and *Pantoea citrea*. Some pests that commonly affect pineapple plants are scales, thrips, mites, mealybugs, ants, and symphylids.

Storage and Transport

Some buyers prefer green fruit, others ripened or off-green. A plant growth regulator, Ethephon, is typically sprayed onto the fruit one week before harvest, developing ethylene, which turns the fruit golden yellow. After cleaning and slicing, a pineapple is typically canned in sugar syrup with added preservative. A pineapple will never become any riper than it was when harvested, though a fully ripe pineapple can bruise and rot quickly. The fruit itself is quite perishable and storage of it should be taken seriously. If it is stored at room temperature, it should be used within two days; however, if it is refrigerated, the time span is extended to five to seven days.

4

Spices, Arils and Dried Flower Buds

A spice is a dried seed, fruit, root, bark, or vegetable substance primarily used for flavoring, colouring or preserving food. Sometimes a spice is used to hide other flavors.

Spices are distinguished from herbs, which are parts of leafy green plants also used for flavoring or as garnish.

Figure: *A group of Indian spices and herbs in bowls*

Many spices have antimicrobial properties. This may explain why spices are more commonly used in warmer climates, which have more infectious disease, and why use of spices is especially prominent in meat, which is particularly susceptible to spoiling.

A spice may have other uses, including medicinal, religious ritual, cosmetics or perfume production, or as a vegetable. For example, turmeric roots are consumed as a vegetable and garlic as an antibiotic.

Early History

Figure: *Spices at central market in Agadir, Morocco*

The spice trade developed throughout South Asia and Middle East in around 2000 BCE with cinnamon and pepper, and in East Asia with herbs and pepper. The Egyptians used herbs for embalming and their demand for exotic herbs helped stimulate world trade. The word *spice* comes from the Old French word *espice*, which became *epice*, and which came from the Latin root *spec*, the noun referring to "appearance, sort, kind": *species* has the same root. By 1000 BCE, medical systems based upon herbs could be found in China, Korea, and India. Early uses were connected with magic, medicine, religion, tradition, and preservation.

Archaeological excavations have uncovered clove burnt onto the floor of a kitchen, dated to 1700 BCE, at the Mesopotamian site of Terqa, in modern-day Syria. The ancient Indian epic Ramayana mentions cloves. The Romans had cloves in the 1st century CE, as Pliny the Elder wrote about them. In the story of Genesis, Joseph was sold into slavery by his brothers to spice merchants. In the biblical poem Song of Solomon, the male speaker compares his beloved to many forms of spices. Generally, early Egyptian, Chinese, Indian, and Mesopotamian sources do not refer to known spices.

Historians believe that nutmeg, which originates from the Banda Islands in South Asia, was introduced to Europe in the 6th century BCE.

Indonesian merchants travelled around China, India, the Middle East, and the east coast of Africa. Arab merchants facilitated the routes through the Middle East and India. This resulted in the Egyptian port city of Alexandria being the main trading centre for spices. The most important discovery prior to the European spice trade were the monsoon winds (40 CE). Sailing from Eastern spice growers to Western European consumers gradually replaced the land-locked spice routes once facilitated by the Middle East Arab caravans.

Middle Ages

Spices were among the most demanded and expensive products available in Europe in the Middle Ages, the most common being black pepper, cinnamon (and the cheaper alternative cassia), cumin, nutmeg, ginger and cloves. Given medieval medicine's main theory of humorism, spices and herbs were indispensable to balance "humors" in food, a daily basis for good health at a time of recurrent pandemics.

Spices were all imported from plantations in Asia and Africa, which made them expensive. From the 8th until the 15th century, the Republic of Venice had the monopoly on spice trade with the Middle East, and along with it the neighbouring Italian city-states. The trade made the region rich. It has been estimated that around 1,000 tons of pepper and 1,000 tons of the other common spices were imported into Western Europe each year during the Late Middle Ages. The value of these goods was the equivalent of a yearly supply of grain for 1.5 million people. The most exclusive was saffron, used as much for its vivid yellow-red colour as for its flavour. Spices that have now fallen into obscurity in European cuisine include grains of paradise, a relative of cardamom which most replaced pepper in late medieval north French cooking, long pepper, mace, spikenard, galangal and cubeb.

Early Modern Period

The control of trade routes and the spice-producing regions were the main reasons that Portuguese navigator Vasco da Gama sailed to India in 1499. Spain and Portugal were not happy to pay the high price that Venice demanded for spices. At around the same time, Christopher Columbus returned from the New World, he described to investors new spices available there.

Another source of competition in the spice trade during the 15th and 16th century was the Ragusans from the maritime republic of Dubrovnik in southern Croatia.

The military prowess of Afonso de Albuquerque (1453–1515) allowed the Portuguese to take control of the sea routes to India. In 1506, he took the island of Socotra in the mouth of the Red Sea and, in 1507, Ormuz in the Persian Gulf. Since becoming the viceroy of the Indies, he took Goa in India in 1510, and Malacca on the Malay peninsula in 1511. The Portuguese could now trade directly with Siam, China, and the Maluku Islands. The Silk Road complemented the Portuguese sea routes, and brought the treasures of the Orient to Europe via Lisbon, including many spices.

With the discovery of the New World came new spices, including allspice, bell and chili peppers, vanilla, and chocolate. This development kept the spice trade, with America as a late comer with its new seasonings, profitable well into the 19th century.

Handling Spices

A spice may be available in several forms: fresh, whole dried, or pre-ground dried. Generally, spices are dried. A whole dried spice has the longest shelf life, so it can be purchased and stored in larger amounts, making it cheaper on a per-serving basis. Some spices are not always available either fresh or whole, for example turmeric, and must commonly, be purchased in ground form. Small seeds, such as fennel and mustard seeds, are often used both whole and in powder form.

The flavour of a spice is derived in part from compounds (volatile oils) that oxidize or evaporate when exposed to air. Grinding a spice greatly increases its surface area and so increases the rates of oxidation and evaporation. Thus, flavour is maximized by storing a spice whole and grinding when needed. The shelf life of a whole dry spice is roughly two years; of a ground spice roughly six months. The "flavour life" of a ground spice can be much shorter. Ground spices are better stored away from light.

To grind a whole spice, the classic tool is mortar and pestle. Less labour-intensive tools are more common now: a microplane or fine grater can be used to grind small amounts; a coffee grinder is useful for larger amounts. A frequently used spice such as black pepper may merit storage in its own hand grinder or mill.

Some flavour elements in spices are soluble in water; many are soluble in oil or fat. As a general rule, the flavors from a spice take time to infuse into the food so spices are added early in preparation.

Salmonella Contamination

Nutrition: Because they tend to have strong flavors and are used in small quantities, spices tend to add few calories to food, even though many spices, especially those made from seeds, contain high portions of fat, protein, and carbohydrate by weight. Many spices, however, can contribute significant portions of micronutrients to the diet. For example, a teaspoon of paprika contains about 1133 IU of Vitamin A, which is over 20% of the recommended daily allowance specified by the US FDA. When used in larger quantity, spices can also contribute a substantial amount of minerals, including iron, magnesium, calcium, and many others, to the diet.

Most herbs and spices have substantial antioxidant activity, owing primarily to phenolic compounds, especially flavonoids, which influence nutrition through many pathways, including affecting the absorption of other nutrients. One study found cumin and fresh ginger to be highest in antioxidant activity. These antioxidants also can act as natural preservatives, preventing or slowing the spoilage of food, leading to a higher nutritional content in stored food. India contributes 70% of global spice production.

Fenugreek

Figure: *Fenugreek (*Trigonella foenum-graecum*) is an annual plant in the family Fabaceae with leaves consisting of three small obovate to oblong leaflets. It is cultivated worldwide as a semi-arid crop, and its seeds are a common ingredient in dishes from the Indian Subcontinent.*

It is believed fenugreek was brought into cultivation in the Near East. While Zohary and Hopf are uncertain which wild strain of the genus *Trigonella* gave rise to domesticated fenugreek, charred fenugreek seeds have been recovered from Tell Halal, Iraq, (carbon dated to 4000 BC) and Bronze Age levels of Lachish and desiccated seeds from the tomb of Tutankhamen. Cato the Elder lists fenugreek with clover and vetch as crops grown to feed cattle.

Production

Major fenugreek-producing countries are Afghanistan, Pakistan, India, Iran, Nepal, Bangladesh, Argentina, Egypt, France, Spain, Turkey and Morocco. The largest producer is India, where the major producing states are Rajasthan, Gujarat, Uttarakhand, Uttar Pradesh, Madhya Pradesh, Maharashtra, Haryana, and Punjab. Rajasthan accounts for over 80% of India's output.

Use

Cuisine: Fenugreek is used as an herb (dried or fresh leaves), spice (seeds), and vegetable (fresh leaves, sprouts, and microgreens). Sotolon is the chemical responsible for fenugreek's distinctive sweet smell.

Cuboid-shaped, yellow-to-amber coloured fenugreek seeds are frequently encountered in the cuisines of the Indian subcontinent, used both whole and powdered in the preparation of pickles, vegetable dishes, daals, and spice mixes such as panch phoron and sambar powder. They are often roasted to reduce bitterness and enhance flavour.

Fresh fenugreek leaves are an ingredient in some Indian curries. Sprouted seeds and microgreens are used in salads. When harvested as microgreens, fenugreek is known as Samudra Methi in Maharashtra, especially in and around Mumbai, where it is often grown near the sea in the sandy tracts, hence the name (Samudra, "ocean" in Sanskrit). *Samudra Methi* is also grown in dry river beds in the Gangetic plains. When sold as a vegetable in India the young plants are harvested with their roots still attached and sold in small bundles in the markets and bazaars. Any remaining soil is washed off to extend their shelf life.

In Persian cuisine, fenugreek leaves are called shanbalile. They are the key ingredient and one of several greens incorporated into ghormeh sabzi and Eshkeneh, often said to be the Iranian national dishes.

Fenugreek is used in Eritrean and Ethiopian cuisine. The word for fenugreek in Amharic is *abesh* (or *abish*), and the seed is used in Ethiopia as a natural herbal medicine in the treatment of diabetes.

Yemenite Jews following the interpretation of Rabbi Salomon Isaacides, Rashi, believe fenugreek, which they call hilbeh, hilba, helba, or halba is the Talmudic Rubia. They use it to produce a sauce also called hilbeh, reminiscent of curry. It is consumed daily and ceremoniously during the meal of the first and/or second night of Rosh Hashana (Jewish New Year).

Nutritional Profile

Fenugreek leaves contain the following nutrients per 100 g of edible portion:

- Carbohydrates: 6.0 g
- Protein: 4.4 g
- Fat: 0.9 g
- Minerals: 1.5 g
- Calcium: 395 mg
- Phosphorus: 51 mg
- Iron: 1.93 mg
- Total energy: 49 kcal

Fennel

Fennel (*Foeniculum vulgare*) is a flowering plant species in the celery family Apiaceae or Umbelliferae. It is treated as the sole species in the genus *Foeniculum* by most botanists. It is a hardy, perennial, herb, with yellow flowers and feathery leaves. It is indigenous to the shores of the Mediterranean but has become widely naturalized in many parts of the world, especially on dry soils near the sea-coast and on riverbanks. It is a highly aromatic and flavorful herb with culinary and medicinal uses and, along with the similar-tasting anise, is one of the primary ingredients of absinthe. Florence fennel or finocchio is a selection with a swollen, bulb-like stem base that is used as a vegetable.

Fennel is used as a food plant by the larvae of some Lepidoptera species including the mouse moth and the anise swallowtail.

Cultural References

As Old English *finule*, fennel is one of the nine plants invoked in the pagan Anglo-Saxon *Nine Herbs Charm*, recorded in the 10th century.

In Greek mythology, Prometheus used the stalk of a fennel plant to steal fire from the demigods. Also, it was from the giant fennel, *Ferula communis*, that the Bacchanalian wands of the demigod Dionysus and his followers were said to have come.

The Greek name for fennel is marathon or *marathos*, and the place of the famous battle of Marathon (whence Marathon, the subsequent sports event), literally means a plain with fennels. The name Funchal, was applied by the first settlers that landed on its shores due to the abundance of wild fennel where, as tradition goes, the primitive burg was built. From the Portuguese word *"funcho"* (*fennel*) and the suffix *"-al"*, to denote *"a plantation of fennel"*:

> *"...Funchal, to whom the captain gave this name, because it was founded in a beautiful forested valley, full of fennel up to the sea..."*

Appearance

Fennel, *Foeniculum vulgare*, is a perennial herb. It is erect, glaucous green, and grows to heights of up to 2.5 m, with hollow stems. The leaves grow up to 40 cm long; they are finely dissected, with the ultimate segments filiform (threadlike), about 0.5 mm wide. (Its leaves are similar to those of dill, but thinner.) The flowers are produced in terminal compound umbels 5–15 cm wide, each umbel section having 20–50 tiny yellow flowers on short pedicels. The fruit is a dry seed from 4–10 mm long, half as wide or less, and grooved.

Cultivation and Uses

Fennel is widely cultivated, both in its native range and elsewhere, for its edible, strongly flavoured leaves and fruits. Its aniseed flavour comes from anethole, an aromatic compound also found in anise and star anise, and its taste and aroma are similar to theirs, though usually not as strong.

The Florence fennel (*Foeniculum vulgare* Azoricum Group; syn. *F. vulgare* var. *azoricum*) is a cultivar group with inflated leaf bases which form a bulb-like structure. It is of cultivated origin, and has a mild anise-like flavour, but is more aromatic and sweeter. Florence fennel plants are smaller than the wild type. Their inflated leaf bases are eaten as a vegetable, both raw and cooked. There are several cultivars of Florence fennel, which is also known by several other names, notably the Italian name *finocchio*. In North American supermarkets, it is often mislabelled as "anise".

Foeniculum vulgare 'Purpureum' or 'Nigra', "bronze-leaved" fennel, is widely available as a decorative garden plant.

Fennel has become naturalised along roadsides, in pastures, and in other open sites in many regions, including northern Europe, the United States, southern Canada and in much of Asia and Australia. It propagates well by seed, and is considered an invasive species and a weed in Australia and the United States. In western North America, fennel can be found from the coastal and inland wildland-urban interface east into hill and mountain areas, excluding desert habitats.

Florence fennel is one of the three main herbs used in the preparation of absinthe, an alcoholic mixture which originated as a medicinal elixir in Switzerland and became, by the late 19th century, a popular alcoholic drink in France and other countries.

Culinary Uses

The bulb, foliage, and seeds of the fennel plant are widely used in many of the culinary traditions of the world. The small flowers of wild fennel (mistakenly known in America as fennel "pollen") are the most potent form of fennel, but also the most expensive. Dried fennel seed is an aromatic, anise-flavoured spice, brown or green in colour when fresh, slowly turning a dull grey as the seed ages. For cooking, green seeds are optimal. The leaves are delicately flavoured and similar in shape to those of dill. The bulb is a crisp vegetable that can be sautéed, stewed, braised, grilled, or eaten raw. They are used for garnishes and to add flavor to salads. They are also added to sauces and served with pudding. The leaves used in soups and fish sauce and sometimes eaten raw as salad.

Fennel seeds are sometimes confused with those of anise, which are similar in taste and appearance, though smaller. Fennel is also used as a flavouring in some natural toothpastes. The seeds are used in cookery and sweet desserts.

Many cultures in India, Pakistan, Afghanistan, Iran and the Middle East use fennel seed in their cookery. It is one of the most important spices in Kashmiri Pandit and Gujarati cooking. It is an essential ingredient of the Assamese/Bengali/Oriya spice mixture *panch phoron* and in Chinese five-spice powders. In many parts of India and Pakistan, roasted fennel seeds are consumed as *mukhwas*, an after-meal digestive and breath freshener. Fennel leaves are used as leafy green vegetables either by themselves or mixed with other vegetables, cooked to be served and consumed as part of a meal, in some parts of India. In Syria and Lebanon, it is used to make a special kind of egg omelette (along with onions, and flour) called *ijjeh*.

Many egg, fish, and other dishes employ fresh or dried fennel leaves. Florence fennel is a key ingredient in some Italian and German salads, often tossed with chicory and avocado, or it can be braised and served as a warm side dish. It may be blanched or marinated, or cooked in risotto.

Fennel seeds are the primary flavor component in Italian sausage.

In Spain the stems of the fennel plant are used in the preparation of pickled eggplants, "berenjenas de Almagro". An herbal tea or tisane can be made from fennel.

Medicinal Uses

Fennel contains anethole, which can explain some of its medical effects: It, or its polymers, act as phytoestrogens. The essence of fennel can be used as a safe and effective herbal drug for primary dysmenorrhea, but could have lower potency than mefenamic acid at the current study level.

Intestinal Tract

Fennel is widely employed as a carminative, both in humans and in veterinary medicine (e.g., dogs), to treat flatulence by encouraging the expulsion of intestinal gas. Anethole is responsible for the carminative action. It also serves as a mild laxative.

On account of its aromatic and carminative properties, fennel fruit is chiefly used medicinally with purgatives to allay their tendency to griping and for this purpose forms one of the ingredients of the well-known compound Liquorice Powder. Fennel water has properties similar to those of anise and dill water: mixed with sodium bicarbonate and syrup, these waters constitute the domestic 'gripe water', used to correct the flatulence of infants. Volatile oil of Fennel has these properties in concentration. Fennel tea, formerly also employed as a carminative, is made by pouring half a pint of boiling water on a teaspoonful of bruised fennel seeds. Fennel can be made into a syrup to treat babies with colic (formerly thought to be due to digestive upset), but long-term ingestion of fennel preparations by babies is a known cause of thelarche.

Eyes

In the Indian subcontinent, fennel seeds are also eaten raw, sometimes with some sweetener, as they are said to improve eyesight. Ancient Romans regarded fennel as the herb of sight. Root extracts were often used in tonics to clear cloudy eyes. Extracts of fennel seed have been shown in animal studies to have a potential use in the treatment of glaucoma.

Blood and Urine

Fennel may be an effective diuretic and a potential drug for treatment of hypertension.

Breastmilk

There are historical anecdotes that fennel is a galactagogue, improving the milk supply of a breastfeeding mother. This use, although not supported by direct evidence, is sometimes justified by the fact that fennel is a source of phytoestrogens, which promote growth of breast tissue. However, normal lactation does not involve growth of breast tissue. Two case reports resulted in illness for the newborn child: "Both mothers had both been drinking more than 2 litres daily of an herbal tea mixture reportedly containing licorice, fennel, anise, and goat's rue " "The authors attributed the maternal and infant symptoms to anethole, which is found in both fennel and anise; however, the anethole levels were not measured in breastmilk, nor were the teas tested for their content."

Other Uses

Syrup prepared from fennel juice was formerly given for chronic coughs. It is one of the plants which is said to be disliked by fleas, and powdered fennel has the effect of driving away fleas from kennels and stables.

India is the leader in production of anise, badian (star anise), fennel and coriander.

Mustard

Mustard is a condiment made from the seeds of a mustard plant (white or yellow mustard, *Sinapis hirta*; brown or Indian mustard, *Brassica juncea*; or black mustard, *B. nigra*). The whole, ground, cracked, or bruised mustard seeds are mixed with water, salt, lemon juice, or other liquids, and sometimes other flavorings and spices, to create a paste or sauce ranging in colour from bright yellow to dark brown.

Commonly paired with meats and cheeses, mustard is a popular addition to sandwiches, salads, hamburgers, and hot dogs. It is also used as an ingredient in many dressings, glazes, sauces, soups, and marinades; as a cream or a seed, mustard is used in the cuisine of India, the Mediterranean, northern and southeastern Europe, Asia, the Americas, and Africa, making it one of the most popular and widely used spices and condiments in the world.

Flavors

The mustard plant ingredient itself has a sharp, hot, pungent flavor.

Mixing ground mustard seeds with water causes a chemical reaction between two compounds in the seed: the enzyme myrosinase and various glucosinolates such as sinigrin, myrosin, and sinalbin. The myrosinase enzyme turns the glucosinolates into various isothiocyanate compounds known generally as mustard oil. The concentrations of different glucosinolates in mustard plant varieties, and the different isothiocyanates that are produced, make different flavors and intensities.

- allyl isothiocyanate and 4-hydroxybenzyl isothiocyanate are responsible for the sharp hot pungent sensation in mustards and in horseradish, wasabi, and garlic. This is because it stimulates the heat and acidity sensing TRPV ion channel TRPV1 on nociceptors (pain sensing nerve cells) in the mouth and nasal passages. The heat of prepared mustard can dissipate with time. This is due to gradual chemical break-up of 4-hydroxybenzyl isothiocyanate.
- Sulforaphane, Phenethyl isothiocyanate, Benzyl isothiocyanate create milder and less pungent intensities and flavors as when found in broccoli, brussels sprouts, water cress, and cabbages.
- The sulfoxide unit in sulforaphane is structurally similar to a thiol which yields onion or garlic-like odors.

Prepared mustard condiment may also have ingredients giving salt, sour (vinegar), and sweet flavors. Turmeric is often added to commercially-prepared mustards, mainly to give them a yellow colour.

Mustard Seed

Mustard seeds are the small round seeds of various mustard plants. The seeds are usually about 1 or 2 mm in diametre. Mustard seeds may be coloured from yellowish white to black. They are important spices in many regional foods. The seeds can come from three different plants: black mustard (*Brassica nigra*), brown Indian mustard (*B. juncea*), and white mustard (*B. hirta/Sinapis alba*).

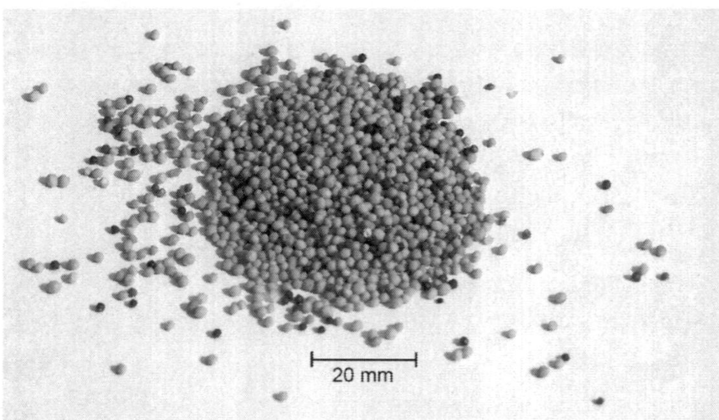

Cultivation

Mustard seeds generally take three to ten days to germinate if placed under the proper conditions, which include a cold atmosphere and relatively moist soil. Mature mustard plants grow into shrubs. Mustard grows well in temperate regions. Major producers of mustard seeds include Canada, Hungary, Great Britain, India, Pakistan and the United States. Brown and black mustard seeds return higher yields than their yellow counterparts.

In Pakistan, rapeseed-mustard is the second most important source of oil, after cotton. It is cultivated over an area of 307,000 hectares with annual production of 233,000 tonnes and contributes about 17% to the domestic production of edible oil.

Mustard seed is a rich source of oil and protein. The seed has oil as high as 46-48 percent, whole seed meal has 43.6 percent protein.

Black Pepper

Black pepper (*Piper nigrum*) is a flowering vine in the family Piperaceae, cultivated for its fruit, which is usually dried and used as a spice and seasoning. The fruit, known as a peppercorn when dried, is approximately 5 millimetres (0.20 in) in diametre, dark red when fully mature, and, like all drupes, contains a single seed. Peppercorns, and the ground pepper derived from them, may be described simply as pepper, or more precisely as black pepper (cooked and dried unripe fruit), green pepper (dried unripe fruit) and white pepper (unripe fruit seeds).

Black pepper is native to south India, and is extensively cultivated there and elsewhere in tropical regions. Currently Vietnam is the world's largest producer and exporter of pepper, producing 34% of the world's *Piper nigrum* crop as of 2008.

Dried ground pepper has been used since antiquity for both its flavor and as a medicine. Black pepper is the world's most traded spice. It is one of the most common spices added to European cuisine and its descendants. The spiciness of black pepper is due to the chemical piperine, not to be confused with the capsaicin that gives fleshy peppers theirs. It is ubiquitous in the modern world as a seasoning, and is often paired with salt.

Varieties

Black pepper: Black pepper is produced from the still-green unripe drupes of the pepper plant. The drupes are cooked briefly in hot water, both to clean them and to prepare them for drying. The heat ruptures cell walls in the pepper, speeding the work of browning enzymes during drying. The drupes are dried in the sun or by machine for several days, during which the pepper around the seed shrinks and darkens into a thin, wrinkled black layer. Once dried, the spice is called black peppercorn. On some estates, the berries are separated from the stem by hand and then sun-dried without the boiling process.

Once the peppercorns are dried, pepper spirit & oil can be extracted from the berries by crushing them. Pepper spirit is used in many medicinal and beauty products. Pepper oil is also used as an ayurvedic massage oil and used in certain beauty and herbal treatments.

Figure: *Black and white peppercorns*

White Pepper: White pepper consists of the seed of the pepper plant alone, with the darker-coloured skin of the pepper fruit removed. This is usually accomplished by a process known as retting, where fully ripe red pepper berries are soaked in water for about a week, during which the flesh of the pepper softens and decomposes. Rubbing then removes what remains of the fruit, and the naked seed is dried. Sometimes alternative processes are used for removing the outer pepper from the seed, including removing the outer layer through mechanical, chemical or biological methods.

Ground white pepper is often used in cream sauces, Chinese and Thai cuisine, and dishes like salad, light-coloured sauces and mashed potatoes, where black pepper would visibly stand out. White pepper has a slightly different flavor than black pepper, due to the lack of certain compounds present in the outer fruit layer of the drupe, but not found in the seed. A slightly sweet version of white pepper from India is sometimes called "whilte pepper".

Green Pepper

Green pepper, like black, is made from the unripe drupes. Dried green peppercorns are treated in a way that retains the green colour, such as treatment with sulphur dioxide, canning or freeze-drying. Pickled peppercorns, also green, are unripe drupes preserved in brine or vinegar. Fresh, unpreserved green pepper drupes, largely unknown in the West, are used in some Asian cuisines, particularly Thai cuisine. Their flavor has been described as spicy and fresh, with a bright aroma. They decay quickly if not dried or preserved.

Figure: *Black, green, pink (*Schinus terebinthifolius*), and white peppercorns*

Orange Pepper and Red Pepper

Orange pepper or red pepper usually consists of ripe red pepper drupes preserved in brine and vinegar. Ripe red peppercorns can also be dried using the same colour-preserving techniques used to produce green pepper.

Pink Pepper and Other Plants Used as Pepper

Pink pepper from *Piper nigrum* is distinct from the more-common dried "pink peppercorns", which are actually the fruits of a plant from a different family, the Peruvian pepper tree, *Schinus molle*, or its relative the Brazilian pepper tree, *Schinus terebinthifolius*.

The bark of Drimys winteri (*Canelo* or *Winter's Bark*) is used as a substitute for pepper in cold and temperate regions of Chile and Argentina where it is easily available.

In New Zealand the seeds of Kawakawa (*Macropiper excelsum*), a relative of black pepper, are sometimes used as pepper and the leaves of *Pseudowintera colourata* (mountain horopito) are another replacement for pepper.

Several plants in the United States are used also as pepper substitutes, such as Lepidium campestre, Lepidium virginicum, shepherd's purse, horseradish,and field Pennycress.

Region of Origin

Peppercorns are often categorized by their place of origin. Two types come from India's Malabar Coast: *Malabar* and *Tellicherry*.

Tellicherry comes from grafted Malabar plants grown on Mount Tellicherry. Sarawak pepper is native to the Malaysian portion of Borneo. White Muntok pepper comes from Indonesia and Lampung hails its island of Sumatra. Vietnam produces both white and black pepper in the provinces of Bà RËa–Ving Tàu, Chu Se and Bình PhýÛc.

Kampot Pepper is native to Kampot, Cambodia. Kampot_(pepper) has received GI (Geographical Indication) in 2008. The pepper grown and designated as Kampot Pepper is grown in a limited geographical region. There are four varieties grown - black, green, red, and white.

Plant

***Figure:** Unripe drupes of Black Pepper (*Piper nigrum*) at Trivandrum, Kerala, India*

The pepper plant is a perennial woody vine growing up to 4 metres (13 ft) in height on supporting trees, poles, or trellises. It is a spreading vine, rooting readily where trailing stems touch the ground. The leaves are alternate, entire, 5 to 10 cm long and 3 to 6 cm across. The flowers are small, produced on pendulous spikes 4 to 8 cm long at the leaf nodes, the spikes lengthening up to 7 to 15 cm as the fruit matures. The fruit of the black pepper is called a drupe and when dried it is a peppercorn.

Pepper can be grown in soil that is neither too dry nor susceptible to flooding, moist, well-drained and rich in organic matter (the vines do not do too well over an altitude of 3000 ft above sea level). The plants are propagated by cuttings about 40 to 50 centimetres long, tied up to neighbouring trees or climbing frames at distances of about two metres apart; trees with rough bark are favoured over those with smooth bark, as the pepper plants climb rough bark more readily. Competing plants are cleared away, leaving only sufficient trees to provide shade and permit free ventilation. The roots are covered in leaf mulch and manure, and the shoots are trimmed twice a year. On dry soils the young plants require watering every other day during the dry season for the first three years. The plants bear fruit from the fourth or fifth year, and typically continue to bear fruit for seven years. The cuttings are usually cultivars, selected both for yield and quality of fruit.

A single stem will bear 20 to 30 fruiting spikes. The harvest begins as soon as one or two fruits at the base of the spikes begin to turn red, and before the fruit is fully mature, and still hard; if allowed to ripen completely, the fruit lose pungency, and ultimately fall off and are lost. The spikes are collected and spread out to dry in the sun, then the peppercorns are stripped off the spikes.

Black pepper is either native to South East Asia. or Southern Asia Within the genus *Piper*, it is most closely related to other Asian species such as *Piper caninum*.

History

Pepper is native to South Asia and Southeast Asia and has been known to Indian cooking since at least 2 BCE. J. Innes Miller notes that while pepper was grown in southern Thailand and in Malaysia, its most important source was India, particularly the Malabar Coast, in what is now the state of Kerala Peppercorns were a much-prized trade good, often referred to as "black gold" and used as a form of commodity money. The legacy of this trade remains in some Western legal systems which recognize the term "peppercorn rent" as a form of a token payment made for something that is in fact being given.

The ancient history of black pepper is often interlinked with (and confused with) that of long pepper, the dried fruit of closely related *Piper longum*. The Romans knew of both and often referred to either as just "piper". In fact, it was not until the discovery of the New World and of chili peppers that the popularity of long pepper entirely declined. Chili peppers, some of which when dried are similar in shape and taste to long pepper, were easier to grow in a variety of locations more convenient to Europe. Before the 16th century, pepper was being grown in Java, Sunda, Sumatra, Madagascar, Malaysia, and everywhere in Southeast Asia. These areas traded mainly with China, or used the pepper locally. Ports in the Malabar area also served as a stop-off point for much of the trade in other spices from farther east in the Indian Ocean. Following the British hegemony in India, virtually all of the black pepper found in Europe, the Middle East, and North Africa was traded from Malabar region.

Black pepper, along with other spices from Southern and Southeast Asia and lands farther east, changed the course of world history. It was in some part the preciousness of these spices that led to the Portuguese efforts to find a sea route to China during the age of discovery and consequently to the Portuguese colonial occupation of that country, as well as the European discovery and colonization of the Americas.

Ancient Times

Black peppercorns were found stuffed in the nostrils of Ramesses II, placed there as part of the mummification rituals shortly after his death in 1213 BCE. Little else is known about the use of pepper in ancient Egypt and how it reached the Nile from Southeast Asia. Pepper (both long and black) was known in Greece at least as early as the 4th century BCE, though it was probably an uncommon and expensive item that only the very rich could afford. Trade routes of the time were by land, or in ships which hugged the coastlines of the Arabian Sea. Long pepper, growing in the north-western part of India, was more accessible than the black pepper from further south; this trade advantage, plus long pepper's greater spiciness, probably made black pepper less popular at the time.

By the time of the early Roman Empire, especially after Rome's conquest of Egypt in 30 BCE, open-ocean crossing of the Arabian Sea direct to southern India's Malabar Coast was near routine. Details of this trading across the Indian Ocean have been passed down in the *Periplus of the Erythraean Sea*. According to the Roman geographer Strabo, the early Empire sent a fleet of around 120 ships on an annual one-year trip to China, Southeast Asia, India and back. The fleet timed its travel across the Arabian Sea to take advantage of the predictable monsoon winds. Returning from India, the ships travelled up the Red Sea, from where the cargo was carried overland or via the Nile Canal to the Nile River, barged to Alexandria, and shipped from there to Italy and Rome. The rough geographical outlines of this same trade route would dominate the pepper trade into Europe for a millennium and a half to come.

With ships sailing directly to the Malabar coast, black pepper was now travelling a shorter trade route than long pepper, and the prices reflected it. Pliny the Elder's *Natural History* tells us the prices in Rome around 77 CE: "Long pepper ... is fifteen denarii per pound, while that of white pepper is seven, and of black, four." Pliny also complains "there is no year in which India does not drain the Roman Empire of fifty million sesterces," and further moralizes on pepper:

It is quite surprising that the use of pepper has come so much into fashion, seeing that in other substances which we use, it is sometimes their sweetness, and sometimes their appearance that has attracted our notice; whereas, pepper has nothing in it that can plead as a recommendation to either fruit or berry, its only desirable quality being a certain pungency; and yet it is for this that we import it all the way from India! Who was the first to make trial of it as an article of food? and who, I wonder, was the man that was not content to prepare himself by hunger only for the satisfying of a greedy appetite? (*N.H.* 12.14)

Black pepper was a well-known and widespread, if expensive, seasoning in the Roman Empire. Apicius' De re coquinaria, a 3rd-century cookbook probably based at least partly on one from the 1st century CE, includes pepper in a majority of its recipes. Edward Gibbon wrote, in *The History of the Decline and Fall of the Roman Empire*, that pepper was "a favourite ingredient of the most expensive Roman cookery".

Postclassical Europe

Pepper was so valuable that it was often used as collateral or even currency. In the Dutch language, "pepper expensive" (*peperduur*) is an expression for something very expensive. The taste for pepper (or the appreciation of its monetary value) was passed on to those who would see Rome fall. Alaric the Visigoth included 3,000 pounds of pepper as part of the ransom he demanded from Rome when he besieged the city in 5th century. After the fall of Rome, others took over the middle legs of the spice trade, first the Persians and then the Arabs; Innes Miller cites the account of Cosmas Indicopleustes, who travelled east to India, as proof that "pepper was still being exported from India in the sixth century". By the end of the Early Middle Ages, the central portions of the spice trade were firmly under Islamic control. Once into the Mediterranean, the trade was largely monopolized by Italian powers, especially Venice and Genoa. The rise of these city-states was funded in large part by the spice trade.

A riddle authored by Saint Aldhelm, a 7th-century Bishop of Sherborne, sheds some light on black pepper's role in England at that time:

I am black on the outside, clad in a wrinkled cover,

Yet within I bear a burning marrow.

I season delicacies, the banquets of kings, and the luxuries of the table,

Both the sauces and the tenderized meats of the kitchen.

But you will find in me no quality of any worth,

Unless your bowels have been rattled by my gleaming marrow.

It is commonly believed that during the Middle Ages, pepper was used to conceal the taste of partially rotten meat. There is no evidence to support this claim, and historians view it as highly unlikely: in the Middle Ages, pepper was a luxury item, affordable only to the wealthy, who certainly had unspoiled meat available as well. In addition, people of the time certainly knew that eating spoiled food would make them sick. Similarly, the belief that pepper was widely used as a preservative is questionable: it is true that piperine, the compound that gives pepper its spiciness, has some antimicrobial properties, but at the concentrations present when pepper is used as a spice, the effect is small. Salt is a much more effective preservative, and salt-cured meats were common fare, especially in winter. However, pepper and other spices certainly played a role in improving the taste of long-preserved meats.

Its exorbitant price during the Middle Ages—and the monopoly on the trade held by Italy—was one of the inducements which led the Portuguese to seek a sea route to India. In 1498, Vasco da Gama became the first person to reach India by sailing around Africa; asked by Arabs in Calicut (who spoke Spanish and Italian) why they had come, his representative replied, "we seek Christians and spices". Though this first trip to India by way of the southern tip of Africa was only a modest success, the Portuguese quickly returned in greater numbers and eventually gained much greater control of trade on the Arabian sea. It was given additional legitimacy (at least from a European imperialistic perspective) by the 1494 Treaty of Tordesillas, which granted Portugal exclusive rights to the half of the world where black pepper originated.

The Portuguese proved unable to maintain their stranglehold on the spice trade for long. The old Arab and Venetian trade networks successfully 'smuggled' enormous quantities of spices through the patchy Portuguese blockade, and pepper once again flowed through Alexandria and Italy, as well as around Africa. In the 17th century, the Portuguese lost almost all of their valuable Indian Ocean trade to the Dutch and the English who, taking advantage from the Spanish ruling over Portugal (1580–1640), occupied by force almost all Portuguese dominations in the area. The pepper ports of Malabar began to trade increasingly with the Dutch in the period 1661–1663.

As pepper supplies into Europe increased, the price of pepper declined (though the total value of the import trade generally did not). Pepper, which in the early Middle Ages had been an item exclusively for the rich, started to become more of an everyday seasoning among those of more average means. Today, pepper accounts for one-fifth of the world's spice trade.

China

It is possible that black pepper was known in China in the 2nd century BCE, if poetic reports regarding an explorer named Tang Meng (U™„) are correct. Sent by Emperor Wu to what is now south-west China, Tang Meng is said to have come across something called *jujiang* or "sauce-betel". He was told it came from the markets of Shu, an area in what is now the Sichuan province. The traditional view among historians is that "sauce-betel" is a sauce made from betel leaves, but arguments have been made that it actually refers to pepper, either long or black.

In the 3rd century CE, black pepper made its first definite appearance in Chinese texts, as *hujiao* or "foreign pepper". It does not appear to have been widely known at the time, failing to appear in a 4th-century work describing a wide variety of spices from beyond China's southern border, including long pepper. By the 12th century, however, black pepper had become a popular ingredient in the cuisine of the wealthy and powerful, sometimes taking the place of China's native Sichuan pepper (the tongue-numbing dried fruit of an unrelated plant).

Marco Polo testifies to pepper's popularity in 13th-century China when he relates what he is told of its consumption in the city of Kinsay (Hangzhou): "... Messer Marco heard it stated by one of the Great Kaan's officers of customs that the quantity of pepper introduced daily for consumption into the city of Kinsay amounted to 43 loads, each load being equal to 223 lbs." Marco Polo is not considered a very reliable source regarding China, and this second-hand data may be even more suspect, but if this estimated 10,000 pounds (4,500 kg) a day for one city is anywhere near the truth, China's pepper imports may have dwarfed Europe's.

During the course of the treasure voyages in the early 15th century, Admiral Zheng He and his expeditionary fleets returned with such a large amount of black pepper that the once-costly luxury became a common commodity.

As Medicine

Like many eastern spices, pepper was historically both a seasoning and a medicine. Long pepper, being stronger, was often the preferred medication, but both were used. Black Pepper (or perhaps long pepper) was believed to cure illness such as constipation, diarrhea, earache, gangrene, heart disease, hernia, hoarseness, indigestion, insect bites, insomnia, joint pain, liver problems, lung disease, oral abscesses, sunburn, tooth decay, and toothaches. Various sources from the 5th century onward also recommend pepper to treat eye problems, often by applying salves or poultices made with pepper directly to the eye. There is no current medical evidence that any of these treatments has any benefit; pepper applied directly to the eye would be quite uncomfortable and possibly damaging. Nevertheless, Black pepper, either powdered or its decoction, is widely used in traditional Indian medicine and as a home remedy for relief from sore throat, throat congestion, cough etc.

Pepper is known to cause sneezing. Some sources say that piperine, a substance present in black pepper, irritates the nostrils, causing the sneezing; Few, if any, controlled studies have been carried out to answer the question. It has been shown that piperine can dramatically increase absorption of selenium, vitamin B, beta-carotene and curcumin as well as other nutrients. As a medicine, pepper appears in the Buddhist Samaññaphala Sutta, chapter five, as one of the few medicines allowed to be carried by a monk.

Pepper contains small amounts of safrole, a mildly carcinogenic compound. Also, it is eliminated from the diet of patients having abdominal surgery and ulcers because of its irritating effect upon the intestines, being replaced by what is referred to as a bland diet. However, extracts from black pepper have been found to have antioxidant properties and anti-carcinogenic effects, especially when compared to chili.

Piperine present in black pepper acts as a thermogenic compound. Piperine enhances the thermogenesis of lipid and accelerates energy metabolism in the body and also increases the serotonin and beta-endorphin production in the brain. Piperine and other components from black pepper may also be helpful in treating vitiligo, although when combined with UV radiation should be staggered due to the effect of light on the compound.

Flavour

Pepper gets its spicy heat mostly from the piperine compound, which is found both in the outer fruit and in the seed. Black pepper contains between 4.6% and 9.7% piperine by mass, and white pepper slightly more than that. Refined piperine, by weight, is about one percent as hot as the capsaicin in chili peppers. The outer fruit layer, left on black pepper, also contains important odour-contributing terpenes including pinene, sabinene, limonene, caryophyllene, and linalool, which give citrusy, woody, and floral notes. These scents are mostly missing in white pepper, which is stripped of the fruit layer. White pepper can gain some different odours (including musty notes) from its longer fermentation stage. The aroma of pepper is attributed to rotundone (3,4,5,6,7,8-Hexahydro-3á,8á-dimethyl-5á-(1-methylethenyl)azulene-1(2H)-one), a sesquiterpene originally discovered in the tubers of cyperus rotundus, which can be detected in concentrations of 0.4 nanograms/L in water and in wine: rotundone is also present in marjoram, oregano, rosemary, basil, thyme, and geranium, as well as in some Shiraz wines.

Pepper loses flavour and aroma through evaporation, so airtight storage helps preserve its spiciness longer. Pepper can also lose flavour when exposed to light, which can transform piperine into nearly tasteless isochavicine. Once ground, pepper's aromatics can evaporate quickly; most culinary sources recommend grinding whole peppercorns immediately before use for this reason. Handheld pepper mills or grinders, which mechanically grind or crush whole peppercorns, are used for this, sometimes instead of pepper shakers that dispense pre-ground pepper. Spice mills such as pepper mills were found in European kitchens as early as the 14th century, but the mortar and pestle used earlier for crushing pepper have remained a popular method for centuries as well.

World Trade

Peppercorns (dried black pepper) are, by monetary value, the most widely traded spice in the world, accounting for 20 percent of all spice imports in 2002. The price of pepper can be volatile, and this figure fluctuates a great deal year to year; for example, pepper made up 39 percent of all spice imports in 1998. By weight, slightly more chili peppers are traded worldwide than peppercorns.

The International Pepper Exchange is located in Kochi, India. Participation in the IPE however is domestic with regulatory restrictions on international membership on local exchanges; something common to almost all Asian commodity exchanges.

As of 2008, Vietnam is the world's largest producer and exporter of pepper, producing 34% of the world's *Piper nigrum*. Other major producers include India (19%), Brazil (13%), Indonesia (9%), Malaysia (8%), Sri Lanka (6%), China (6%), and Thailand (4%). Global pepper production peaked in 2003 with over 355,000 t (391,000 short tons), but has fallen to just over 271,000 t (299,000 short tons) by 2008 due to a series of issues including poor crop management, disease and weather. Vietnam dominates the export market, using almost none of its production domestically; however its 2007 crop fell by nearly 10% from the previous year to about 90,000 t (99,000 short tons).

Similar crop yields occurred in 2007 across the other pepper producing nations as well. Nowadays, in England, industrial buyers mix Peppers of different origin to maintain a balance between price, taste and other factors. Malabar black peppers are used for weight and taste, Sumatra for colour, and Penang for strength.

Arils, Barks and Dried Flower Buds

An aril, also called an arillus, is any specialized outgrowth from the funiculus (attachment point of the seed) (or *hilum*) that covers or is attached to the seed. The term is sometimes applied to any fleshy appendage of the seed in flowering plants, such as the mace of the nutmeg seed. The aril is often an edible enticement, encouraging transport by animals and thereby assisting in seed dispersal. Pseudarils are aril-like structures commonly found on the pyrenes of Burseraceae species that develop from the mesocarp of the ovary – the fleshy, edible pericarp splits neatly in two halves, then falling away or being eaten to reveal a brightly coloured pseudaril around the black seed.

The aril may create a fruit-like structure (called a *false-fruit*). False fruit are found in numerous Angiosperm taxa. The edible flesh of the longan, lychee, ackee and lleuque fruits are highly developed arils surrounding the seed rather than a pericarp layer. Such arils are also found in a few species of gymnosperms, notably the yews and related conifers. Instead of the woody cone typical of most gymnosperms, the reproductive structure of the yew consists of a single seed that becomes surrounded by a fleshy, cup-like covering. This covering is derived from a highly modified cone scale.

Taxus Baccata

Taxus baccata is a conifer native to western, central and southern Europe, northwest Africa, northern Iran and southwest Asia. It is the tree originally known as yew, though with other related trees becoming known, it may now be known as English yew, or European yew.

The common yew was one of the many species first described by Linnaeus. Along with around 30 other species, it is classified in the family Taxaceae, which is now firmly classified as a conifer in the order Pinales.

Figure: Taxus baccata *(European yew) shoot with mature and immature cones*

Description

It is a small- to medium-sized evergreen tree, growing 10–20 metres (33–66 ft) (exceptionally up to 28 metres (92 ft)) tall, with a trunk up to 2 metres (6 ft 7 in) (exceptionally 4 metres (13 ft)) diametre. The bark is thin, scaly brown, coming off in small flakes aligned with the stem. The leaves are flat, dark green, 1–4 centimetres (0.39–1.57 in) long and 2–3 millimetres (0.079–0.118 in) broad, arranged spirally on the stem, but with the leaf bases twisted to align the leaves in two flat rows either side of the stem, except on erect leading shoots where the spiral arrangement is more obvious. The leaves are highly poisonous.

The seed cones are highly modified, each cone containing a single seed 4–7 millimetres (0.16–0.28 in) long partly surrounded by a modified scale which develops into a soft, bright red berry-like structure called an aril, 8–15 millimetres (0.31–0.59 in) long and wide and open at the end. The arils mature 6 to 9 months after pollination, and with the seed contained are eaten by thrushes, waxwings and other birds, which disperse the hard seeds undamaged in their droppings; maturation of the arils is spread over 2 to 3 months, increasing the chances of successful seed dispersal.

The seeds themselves are extremely poisonous and bitter, but are opened and eaten by some bird species including Hawfinches Greenfinches and Great Tits. The aril is not poisonous, and is gelatinous and very sweet tasting. The male cones are globose, 3–6 millimetres (0.12–0.24 in) diametre, and shed their pollen in early spring. It is mostly dioecious, but occasional individuals can be variably monoecious, or change sex with time.

Longevity

Taxus baccata can reach 400 to 600 years of age. Some specimens live longer but the age of yews is often overestimated. Ten yews in Britain are believed to predate the 10th century. The potential age of yews is impossible to determine accurately and is subject to much dispute. There is rarely any wood as old as the entire tree, while the boughs themselves often become hollow with age, making ring counts impossible. There are claims as high as 5,000–9,500 years, but other evidence based on growth rates and archaeological work of surrounding structures suggests the oldest trees (such as the Fortingall Yew in Perthshire, Scotland) are more likely to be in the range of 2,000 years. Even with this lower estimate, *Taxus baccata* is one of the longest-living plants in Europe. One characteristic contributing to its longevity is that it is able to split under the weight of advanced growth without succumbing to disease in the fracture, as do most other trees. Another is its ability to give rise to new epicormic and basal shoots from cut surfaces and low on its trunk, even at an old age.

Significant Trees

The Fortingall Yew in Perthshire, Scotland, has the largest recorded trunk girth in Britain and experts estimate it to be 2,000 to 3,000 years old, although it may be a remnant of a post-Roman Christian site and around 1,500 years old. The Llangernyw Yew in Clwyd, Wales, can be found at an early saint site and is about 1,500 years old. Other well known yews include the Ankerwycke Yew, the Balderschwang Yew, the Caesarsboom, the Florencecourt Yew, and the Borrowdale Fraternal Four, of which poet William Wordsworth wrote. The Kingley Vale National Nature Reserve in West Sussex has one of Europe's largest yew woodlands.

The oldest specimen in Spain is located in Bermiego, Asturias. It is known as *Teixu l'Iglesia* in the Asturian language. It stands 15 m (49 ft) tall with a trunk diametre of 6.82 m (22.4 ft) and a crown diametre of 15 m. It was declared a Natural Monument on April 27, 1995 by the Asturian Government and is protected by the Plan of Natural Resources.

A unique forest formed by *Taxus baccata* and European Box (*Buxus sempervirens*) lies within the city of Sochi, in the Western Caucasus.

Toxicity

Most parts of the tree are toxic, except the bright red aril surrounding the seed. The foliage remains toxic even when wilted, and toxicity increases in potency when dried. Ingestion and subsequent excretion by birds whose beaks and digestive systems do not break down the seed's coating are the primary means of yew dispersal. The major toxin within the yew is the alkaloid taxine. Horses have a relatively low tolerance to taxine, with a lethal dose of 200–400 mg/kg body weight; cattle, pigs, and other livestock are only slightly less vulnerable. Several studies have found taxine LD_{50} values under 20 mg/kg in mice and rats.

Symptoms of yew poisoning include an accelerated heart rate, muscle tremors, convulsions, collapse, difficulty breathing, circulation impairment and eventually heart failure. However, there may be no symptoms, and if poisoning remains undetected death may occur within hours. Fatal poisoning in humans is very rare, usually occurring after consuming yew foliage. The leaves are more toxic than the seed.

Uses and Traditions

One of the world's oldest surviving wooden artifacts is a Clactonian yew spear head, found in 1911 at Clacton-on-Sea, in Essex, UK. It is estimated to be about 450,000 years old.

In the ancient Celtic world, the yew tree (**eburos*) had extraordinary importance; a passage by Caesar narrates that Catuvolcus, chief of the Eburones poisoned himself with yew rather than submit to Rome (*Gallic Wars* 6: 31). Similarly, Florus notes that when the Cantabrians were under siege by the legate Gaius Furnius in 22 BC, most of them took their lives either by the sword, by fire, or by a poison extracted *ex arboribus taxeis*, that is, from the yew tree (2: 33, 50–51). In a similar way, Orosius notes that when the Astures were besieged at *Mons Medullius*, they preferred to die by their own swords or by the yew tree poison rather than surrender (6, 21, 1).

Religion

The yew is often found in churchyards in England, Wales, Scotland, Ireland, France and northern areas of Spain. In France, the oldest yew trees are almost all located in church yards of Normandy and a chapel was very often laid out in the hollow trunk. Some examples can be found in La Haye-de-Routot or La Lande-Patry.

It is said that up to 40 people could stand inside one of the La-Haye-de-Routot yew trees and the Le Ménil-Ciboult yew is probably the largest one (13 m diametre). Indeed some of these trees are exceptionally large (over 5 m diametre) and may be over 2,000 years old. Sometimes monks planted yews in the middle of their cloister, as at Muckross Abbey (Ireland) or abbaye de Jumièges (France). Some ancient yew trees are located at St Mary the Virgin Church, Overton-on-Dee in Wales.

In Asturian tradition and culture the yew tree has had a real link with the land, the people, the ancestors and the ancient religion. It was tradition on All Saints Day to bring a branch of a yew tree to the tombs of those who had died recently so they will find the guide in their return to the Land of Shadows.The yew tree can be found near chapels, churches and cemetreies since ancient times as a symbol of the transcendence of death, and is usually found in the main squares of the villages where people celebrated the open councils that served as a way of general assembly to rule the village affairs.

It has been suggested that the Sacred Tree at the Temple at Uppsala was an ancient yew tree. The Christian church commonly found it expedient to take over existing pre-Christian sacred sites for churches. It has also been suggested that yews were planted at religious sites as their long life was suggestive of eternity, or because being toxic they were seen as trees of death. Another suggested explanation is that yews were planted to discourage farmers and drovers from letting animals wander onto the burial grounds, the poisonous foliage being the disincentive. A further possible reason is that fronds and branches of yew were often used as a substitute for palms on Palm Sunday.

In traditional Germanic religion was Yggdrasill often seen as a giant ash tree. Many scholars are now agree that in the past an error has been made in the interpretation of the ancient writings, and that the tree is most likely an European yew (Taxus baccata). This mistake would find its origin in an alternative word for the yew tree in the Old Norse, namely Needle Ash (barraskr). In addition, ancient sources, including the Eddas, speak about a *vetgrønster vida* which means "evergreen tree". An ash sheds its leaves in the winter, while yew trees retain their needles.

Conifers were in the past often seen as sacred, because they never lose their green. In addition, the tree of life was not only an object from the stories, but also believers often gathered around an existing tree. The yew releases gaseous toxins (taxine) on hot days. Taxine is in some instances capable of causing hallucinations. This has some similarities with the story that Odin had an revelation (the wisdom of the runes) after having been hanging from the tree for nine days.

Medicines

In 1021, Avicenna introduced the medicinal use of *T. baccata* for phytotherapy in *The Canon of Medicine*. He named this herbal drug "Zarnab" and used it as a cardiac remedy. This was the first known use of a calcium channel blocker drug, which were not in wide use in the Western world until the 1960s.

Certain compounds found in the bark of yew trees were discovered by Wall and Wani in 1967 to have efficacy as anti-cancer agents. The precursors of the chemotherapy drug paclitaxel (taxol) can be synthesized easily from the extracts of the leaves of European yew, which is a more renewable source than the bark of the Pacific yew (*Taxus brevifolia*). This ended a point of conflict in the early 1990s; many environmentalists, including Al Gore, had opposed the harvesting of yew for paclitaxel cancer treatments. Docetaxel can then be obtained by semi-synthetic conversion from the precursors.

In the Central Himalayas, the plant is used as a treatment for breast and ovarian cancer.

Woodworking and Longbows

Wood from the yew is classified as a closed-pore softwood, similar to cedar and pine. Easy to work, yew is among the hardest of the softwoods; yet it possesses a remarkable elasticity, making it ideal for products that require springiness, such as bows.

Yew is also associated with Wales and England because of the longbow, an early weapon of war developed in northern Europe, and as the English longbow the basis for a medieval tactical system. The oldest surviving yew longbow was found at Rotten Bottom in Dumfries and Galloway, Scotland. It has been given a calibrated radiocarbon date of 4040 BC to 3640 BC and is on display in the National Museum of Scotland. Yew is the wood of choice for longbow making; the bows are constructed so that the heartwood of yew is on the inside of the bow while the sapwood is on the outside. This takes advantage of the natural properties of yew wood since the heartwood resists compression while the sapwood resists stretching. This increases the strength and efficiency of the bow. Much yew is knotty and twisted, and therefore unsuitable for bowmaking; most trunks do not give good staves and even in a good trunk much wood has to be discarded.

The trade of yew wood to England for longbows was so robust that it depleted the stocks of good-quality, mature yew over a vast area. The first documented import of yew bowstaves to England was in 1294. In 1350 there was a serious shortage, and Henry IV of England ordered his royal bowyer to enter private land and cut yew and other woods. In 1470 compulsory archery practice was renewed, and hazel, ash, and laburnum were specifically allowed for practice bows. Supplies still proved insufficient, until by the Statute of Westminster in 1472, every ship coming to an English port had to bring four bowstaves for every tun. Richard III of England increased this to ten for every tun. This stimulated a vast network of extraction and supply, which formed part of royal monopolies in southern Germany and Austria. In 1483, the price of bowstaves rose from two to eight pounds per hundred, and in 1510 the Venetians would only sell a hundred for sixteen pounds. In 1507 the Holy Roman Emperor asked the Duke of Bavaria to stop cutting yew, but the trade was profitable, and in 1532 the royal monopoly was granted for the usual quantity "if there are that many." In 1562, the Bavarian government sent a long plea to the Holy Roman Emperor asking him to stop the cutting of yew, and outlining the damage done to the forests by its selective extraction, which broke the canopy and allowed wind to destroy neighbouring trees. In 1568, despite a request from Saxony, no royal monopoly was granted because there was no yew to cut, and the next year Bavaria and Austria similarly failed to produce enough yew to justify a royal monopoly. Forestry records in this area in the 17th century do not mention yew, and it seems that no mature trees were to be had. The English tried to obtain supplies from the Baltic, but at this period bows were being replaced by guns in any case.

Horticulture

Today European yew is widely used in landscaping and ornamental horticulture. Due to its dense, dark green, mature foliage, and its tolerance of even very severe pruning, it is used especially for formal hedges and topiary. Its relatively slow growth rate means that in such situations it needs to be clipped only once per year (in late summer).

Well over 200 cultivars of *T. baccata* have been named. The most popular of these are the Irish yew (*T. baccata* 'Fastigiata'), a fastigiate cultivar of the European yew selected from two trees found growing in Ireland, and the several cultivars with yellow leaves, collectively known as "golden yew". In some locations, e.g. when hemmed in by buildings or other trees, an Irish yew can reach 20 feet in height without exceeding 2 feet in diametre at its thickest point, although with age many Irish yews assume a fat cigar shape rather than being truly columnar.

European yew will tolerate growing in a wide range of soils and situations, including shallow chalk soils and shade, although in deep shade its foliage may be less dense. However it cannot tolerate waterlogging, and in poorly-draining situations is liable to succumb to the root-rotting pathogen *Phytophthora cinnamomi*.

In Europe, *Taxus baccata* grows naturally north to Molde in southern Norway, but it is used in gardens further north. It is also popular as a bonsai in many parts of Europe and makes a handsome small to large sized bonsai.

Musical Instruments

The late Robert Lundberg, a noted luthier who performed extensive research on historical lute-making methodology, states in his 2002 book *Historical Lute Construction* that yew was historically a prized wood for lute construction. European legislation establishing use limits and requirements for yew limited supplies available to luthiers, but it was apparently as prized among medieval, renaissance, and baroque lute builders as Brazilian Rosewood is among contemporary guitar-makers for its quality of sound and beauty.

Conservation

Clippings from ancient specimens in the UK, including the Fortingall Yew, were taken to the Royal Botanic Gardens in Edinburgh to form a mile-long hedge. The purpose of this "Yew Conservation Hedge Project" is to maintain the DNA of *Taxus baccata.* The species is threatened by felling, partly due to rising demand from pharmaceutical companies, and disease.

Dacrycarpus Dacrydioides

The kahikatea tree, *Dacrycarpus dacrydioides,* is native to New Zealand. In pre-European times the aril of the kahikatea was a food source for Mâori. The washed arils were called *koroi* and were eaten raw.

Dacrycarpus dacrydioides or kahikatea (from its name in the Mâori language) is a coniferous tree endemic to New Zealand.

The tree grows to a height of 55 metres (180 ft) with a trunk exceeding 1 metre (3 ft) in diametre, and is buttressed at the base. It is dominant in lowland forest and wetlands throughout the North and South Islands. The leaves are spirally arranged; on young plants, they are awl-shaped, 3 to 8 mm long, and twisted at the base to lie spread to the sides of the shoot in a flat plane; on mature trees, they are scale-like, 1 to 3 mm long, and placed all round the shoot. The cones are highly modified, with the cone scales swelling at maturity into an orange to red, fleshy, aril with a single apical seed 3 to 5 mm in diametre. The seeds are dispersed by birds, which eat the fleshy scale and pass the seeds in their droppings.

Before extensive logging, trees of 80 m height were known. A specimen in Pirongia Forest Park in the Waikato region is the tallest native tree in New Zealand at 66.5 metres (218 ft). Tâne Mahuta, the famously large kauri tree (Agathis australis), is 51.2 m.

The tree was previously referred to by the misleading name "white pine", despite its not being a pine; the Mâori name kahikatea is now more widely used (other Mâori names are kaikatea, kahika, katea, kôaka, koroî). Like many other species in the family Podocarpaceae, the classification of kahikatea has changed over time, having also been placed in the genera *Podocarpus* and *Nageia*. The Waikato still has small stands of natural Kahikatea. The most readily accessible is at Claudelands Bush in the Hamilton suburb of Claudelands which was previously a swamp.

Uses

The timber was initially not so highly regarded by European settlers as it was less attractive, strong, or resilient to rot than the other coniferous timbers available such as totara. However, since the wood does not impart an odour, and is clean and lightweight, Kahikatea was used to make boxes for the exporting of butter when the refrigerated export became feasible from Australia and New Zealand in the 1880s. The butter was exported in 56 lb slabs, and kahikatea became less common as the export industry grew. Indeed, tall mature pure kaihikatea forest, a once distinctive and widespread wetland type, is all but gone except for very small patches mostly on the West Coast. The first bridge in Hamilton across the Waikato River was built mainly of Kahikatea about 1880 but the bridge quickly became unsafe due to rot. The piles (not of Kahikatea) are still visible in the vegetation alongside the Traffic (Bridge St.) Bridge.

Kahikatea was used for boat building up until the 1970s because of its long straight lengths. When 12% dry it is slightly lighter in weight than Kauri and white in colour. Kahikatea was also used as late as the 1980s to carve out *waka*, traditional vessels for Maori, in which they competed in various watersports in Kaituna Wetlands.

For Mâori, the kahikatea had many uses. The fleshy aril or *koroi* was an important food resource, and was served at feasts in great amounts. The wood was also favoured for making bird spears. Soot obtained from burning the heartwood supplied a pigment for traditional tattooing (tâ moko).

Kahikatea, along with other trees in privately owned forests, can only be harvested under a permit system and if sustainable harvesting techniques are used. There are still a few old-growth remnants of kahikatea left in the Waikato (for example, at Mangapu and Mokau).

However, most kahikatea stands in the Waikato lowlands are relatively young forest fragments, about 80-100 years old. These fragments have grown up around a few old trees left standing after most of the original forest was cleared for farming. Some stands still have their original seed trees around which the new forest grew. These massive trees, with trunks occasionally over 2 m across, are about 400-500 years old.

There are about 370 kahikatea fragments in the Waikato region. They are typically small, between 0.5 and 50 ha, with half of them less than 5 ha. Most of them grow on the river floodplains of the Waikato Basin, Hauraki Plains and Mokau River.

Because we now control flood events and re-sow flood-damaged pasture, we are unlikely to see many new areas of kahikatea forest developing in the Waikato. Scientists estimate that more than 98 of percent of the pre-European kahikatea forest has been lost nationwide.

Nutmeg (Mace)

The nutmeg tree is any of several species of trees in genus *Myristica*. The most important commercial species is *Myristica fragrans*, an evergreen tree indigenous to the Banda Islands in the Moluccas (or Spice Islands) of Indonesia.

The nutmeg tree is important for two spices derived from the fruit: nutmeg and mace.

Nutmeg is the seed of the tree, roughly egg-shaped and about 20 to 30 mm (0.8 to 1.2 in) long and 15 to 18 mm (0.6 to 0.7 in) wide, and weighing between 5 and 10 g (0.2 and 0.4 oz) dried, while mace is the dried "lacy" reddish covering or aril of the seed. The first harvest of nutmeg trees takes place 7–9 years after planting, and the trees reach full production after twenty years. Nutmeg is usually used in powdered form. This is the only tropical fruit that is the source of two different spices. Several other commercial products are also produced from the trees, including essential oils, extracted oleoresins, and nutmeg butter.

The common or fragrant nutmeg, *Myristica fragrans*, native to the Banda Islands of Indonesia, is also grown in Penang Island in Malaysia and the Caribbean, especially in Grenada. It also grows in Kerala, a state in southern India. Other species of nutmeg include Papuan nutmeg *M. argentea* from New Guinea, and *M. malabarica* from India.

Botany and Cultivation

Nutmeg is a dioecious plant which is propagated sexually and asexually, the latter being the standard. Sexual propagation by seedling yields 50% male seedlings, which are unproductive. As there is no reliable method of determining plant sex before flowering in the sixth to eighth year, and sexual propagation bears inconsistent yields, grafting is the preferred method of propagation. Epicotyl grafting, approach grafting and patch budding have proved successful, epicotyl grafting being the most widely adopted standard. Air-layering, or marcotting, is an alternative, though not preferred, method, because of its low (35-40%) success rate.

Figure: Myristica fragrans *tree in Goa, India*

Figure: *Nutmegs on a tree in Kerala, India*

Culinary Uses

Nutmeg and mace have similar sensory qualities, with nutmeg having a slightly sweeter and mace a more delicate flavour. Mace is often preferred in light dishes for the bright orange, saffron-like hue it imparts. Nutmeg is used for flavouring many dishes, usually in ground or grated form, and is best grated fresh in a nutmeg grater.

In Penang cuisine, dried, shredded nutmeg rind with sugar coating is used as toppings on the uniquely Penang *ais kacang*. Nutmeg rind is also blended (creating a fresh, green, tangy taste and white colour juice) or boiled (resulting in a much sweeter and brown juice) to make iced nutmeg juice.

In Indian cuisine, nutmeg is used in many sweet as well as savoury dishes (predominantly in Mughlai cuisine). It is also added in small quantities as a medicine for infants. It may also be used in small quantities in *garam masala*. Ground nutmeg is also smoked in India.

In Indonesian cuisine, nutmeg is used in various dishes, mainly in many soups, such as soto soup, baso soup or *Sup Kambing*. It is also made as sweets.

In Middle Eastern cuisine, ground nutmeg is often used as a spice for savoury dishes.

In originally European cuisine, nutmeg and mace are used especially in potato dishes and in processed meat products; they are also used in soups, sauces, and baked goods. In Dutch cuisine, nutmeg is added to vegetables such as Brussels sprouts, cauliflower, and string beans. Nutmeg is a traditional ingredient in mulled cider, mulled wine, and eggnog.

Japanese varieties of curry powder include nutmeg as an ingredient.

In the Caribbean, nutmeg is often used in drinks such as the Bushwacker, Painkiller, and Barbados rum punch. Typically, it is just a sprinkle on the top of the drink.

The pericarp (fruit/pod) is used in Grenada and also in Indonesia to make jam, or is finely sliced, cooked with sugar, and crystallised to make a fragrant candy.

In Scotland, mace and nutmeg are usually both essential ingredients in haggis.

In the US, nutmeg is known as the main pumpkin pie spice and often shows up in simple recipes for other winter squashes such as baked acorn squash.

Essential Oils

The essential oil obtained by steam distillation of ground nutmeg is used widely in the perfumery and pharmaceutical industries. This volatile fraction typically contains 60-80% d-camphene by weight, as well as quantities of d-pinene, limonene, d-borneol, l-terpineol, geraniol, safrol, and myristicin. The oil is colourless or light yellow, and smells and tastes of nutmeg. It contains numerous components of interest to the oleochemical industry, and is used as a natural food flavouring in baked goods, syrups, beverages, and sweets. It is used to replace ground nutmeg, as it leaves no particles in the food. The essential oil is also used in the cosmetic and pharmaceutical industries, for instance, in toothpaste, and as a major ingredient in some cough syrups. In traditional medicine, nutmeg and nutmeg oil were used for disorders related to the nervous and digestive systems. After extraction of the essential oil, the remaining seed, containing much less flavour, is called "spent". Spent is often mixed in industrial mills with pure nutmeg to facilitate the milling process, as nutmeg is not easy to mill due to the high percentage of oil in the pure seed. Ground nutmeg with a variable percentage of spent (around 10% w/w) is also less likely to clot. To obtain a better running powder also a small percentage of rice flour can be added.

Nutmeg Butter

Nutmeg butter is obtained from the nut by expression. It is semi-solid, reddish brown in colour, and tastes and smells of nutmeg. Approximately 75% (by weight) of nutmeg butter is trimyristin, which can be turned into myristic acid, a 14-carbon fatty acid, which can be used as a replacement for cocoa butter, can be mixed with other fats like cottonseed oil or palm oil, and has applications as an industrial lubricant.

Figure: *Nutmeg seeds showing "veins"*

Nutmeg is known to have been a prized and costly spice in European medieval cuisine as a flavouring, medicinal, and preservative agent. Saint Theodore the Studite (ca. 758 – ca. 826) allowed his monks to sprinkle nutmeg on their pease pudding when required to eat it. In Elizabethan times, because nutmeg was believed to ward off the plague, demand increased and its price skyrocketed.

Until the mid-19th century the small island of Banda (known to early English adventurers as "Run", although it is a different island, —both are now part of the Banda Islands), was the world's only source of nutmeg and mace. Nutmeg was known as a valuable commodity by Muslim sailors from the port of Basra (including the fictional character Sinbad the Sailor in the *One Thousand and One Nights*). Nutmeg was traded by Arabs during the Middle Ages and sold to the Venetians for high prices, but the traders did not divulge the exact location of their source in the profitable Indian Ocean trade, and no European was able to deduce its location.

***Figure:** Mace (red) within nutmeg fruit*

In August 1511, Afonso de Albuquerque conquered Malacca, which at the time was the hub of Asian trade, on behalf of the king of Portugal. In November of the same year, after having secured Malacca and learning of Banda's location, Albuquerque sent an expedition of three ships led by his friend António de Abreu to find it. Malay pilots, either recruited or forcibly conscripted, guided them via Java, the Lesser Sundas and Ambon to Run, arriving in early 1512.

The first Europeans to reach the Bandas, the expedition remained in Banda for about a month, purchasing and filling their ships with Banda's nutmeg and mace, and with cloves in which Banda had a thriving entrepôt trade. The first written accounts of Banda are in *Suma Oriental*, a book written by the Portuguese apothecary Tomé Pires, based in Malacca from 1512 to 1515. Full control of this trade by the Portuguese was not possible, and they remained participants without a foothold in the islands themselves.

The trade in nutmeg later became dominated by the Dutch in the 17th century. The English and Dutch, through their competing East India and Dutch East India Companies, engaged in prolonged struggles to gain control of Run Island. At the end of the Second Anglo-Dutch War, the Dutch gained control of Run, while England controlled New Amsterdam (New York) in North America.

The Dutch waged a bloody war, including the massacre and enslavement of the inhabitants of the island of Banda, just to control nutmeg production in the East Indies in 1621. Thereafter, the Banda Islands were run as a series of plantation estates, with the Dutch mounting annual expeditions in local war-vessels to extirpate nutmeg trees planted elsewhere.

In 1760, the price of nutmeg in London was 85 to 90 shillings per pound, a price kept artificially high by the Dutch voluntarily burning warehouses full of nutmegs in Amsterdam.

As a result of the Dutch interregnum during the Napoleonic Wars, the British took temporary control of the Banda Islands from the Dutch and transplanted nutmeg trees, complete with soil, to Sri Lanka, Penang, Bencoolen and Singapore. However, there is evidence that the tree existed in Sri Lanka prior to this. Thence they were transplanted to their other colonial holdings elsewhere, notably Zanzibar and Grenada. The national flag of Grenada, adopted in 1974, shows a stylised split-open nutmeg fruit. The Dutch however continued to hold control of the spice islands until World War II.

Connecticut received its nickname ("the Nutmeg State", "Nutmegger") from the legend that some unscrupulous Connecticut traders would whittle "nutmeg" out of wood, creating a "wooden nutmeg", a term which later came to mean any type of fraud.

World Production

World production of nutmeg is estimated to average between 10,000 and 12,000 tonnes (9,800 and 12,000 long tons; 11,000 and 13,000 short tons) per year, with annual world demand estimated at 9,000 tonnes (8,900 long tons; 9,900 short tons); production of mace is estimated at 1,500 to 2,000 tonnes (1,500 to 2,000 long tons; 1,700 to 2,200 short tons). Indonesia and Grenada dominate production and exports of both products, with world market shares of 75% and 20% respectively. Other producers include India, Malaysia (especially Penang, where the trees grow wild within untamed areas), Papua New Guinea, Sri Lanka, and Caribbean islands, such as St. Vincent. The principal import markets are the European Community, the United States, Japan and India. Singapore and the Netherlands are major re-exporters.

Cinnamon

Cinnamon is a spice obtained from the inner bark of several trees from the genus *Cinnamomum* that is used in both sweet and savoury foods. While *Cinnamomum verum* is sometimes considered to be "true cinnamon", most cinnamon in international commerce is derived from related species, which are also referred to as "cassia" to distinguish them from "true cinnamon". Cinnamon is the name for perhaps a dozen species of trees and the commercial spice products that some of them produce. All are members of the genus *Cinnamomum* in the family Lauraceae. Only a few of them are grown commercially for spice.

Cinnamon has been known from remote antiquity. It was imported to Egypt as early as 2000 BC, but those who report that it had come from China confuse it with cassia.

Figure: *Cinnamon sticks*

The Hebrew Bible makes specific mention of the spice many times: first when Moses is commanded to use both sweet cinnamon and cassia in the holy anointing oil; in Proverbs where the lover's bed is perfumed with myrrh, aloes, and cinnamon; and in Song of Solomon, a song describing the beauty of his beloved, cinnamon scents her garments like *the smell of Lebanon.* Cassia was also part of the ketoret, the consecrated incense described in the Hebrew Bible and Talmud. It is also referred to as the HaKetoret (the incense). It was offered on the specialized incense altar in the time when the Tabernacle was located in the First and Second Jerusalem Temples. The ketoret was an important component of the Temple service in Jerusalem. Psalm 45:8 mentions the garments of the king (or of Torah scholars) that smell of myrrh, aloes and cassia.

It was so highly prized among ancient nations that it was regarded as a gift fit for monarchs and even for a god: a fine inscription records the gift of cinnamon and cassia to the temple of Apollo at Miletus.

Though its source was kept mysterious in the Mediterranean world for centuries by the middlemen who handled the spice trade, to protect their monopoly as suppliers, cinnamon is native to Bangladesh, Sri Lanka, the Malabar Coast of India and Burma.

The first Greek reference to *kasia* is found in a poem by Sappho in the seventh century BC. According to Herodotus, both cinnamon and cassia grew in Arabia, together with incense, myrrh, and ladanum, and are guarded by winged serpents. The phoenix was reputed to build its nest from cinnamon and cassia. But Herodotus mentions other writers who see the home of Dionysos, somewhere east or south of Greece, as the source of cassia. While Theophrastus gives a rather good account of the plants, a curious method for harvesting (worms eat away the wood and leave the bark behind) was described.

The Greeks used *kásia* or *malabathron* to flavour wine, together with absinth wormwood (*Artemisia absinthium*). Egyptian recipes for *kyphi*, an aromatic used for burning, included cinnamon and cassia from Hellenistic times onwards. The gifts of Hellenistic rulers to temples sometimes included cassia and cinnamon, as well as incense, myrrh, and Indian incense (*kostos*), so one can conclude that the Greeks used it in this way too.

Pliny gives a fascinating account of the early spice trade across the Red Sea that cost Rome 100 million sesterces each year. The cinnamon was brought around the Arabian peninsula on "rafts without rudders or sails or oars", taking advantage of the winter trade winds. Pliny also mentions cassia as a flavouring agent for wine.

According to Pliny, a pound (the Roman pound, 327 grams (11.5 oz)) of cassia, cinnamon, or serichatum cost up to 300 *denarii*, the wage of ten months' labour. Diocletian's Edict on Maximum Prices from 301 AD gives a price of 125 *denarii* for a pound of cassia, while an agricultural labourer earned 25 *denarii* per day. It was too expensive to be commonly used on funeral pyres in Rome, but the Emperor Nero is said to have burned a year's worth of the city's supply at the funeral for his wife Poppaea Sabina in AD 65.

Malabathrum leaves (*folia*) were used in cooking and for distilling an oil used in a caraway-sauce for oysters by the Roman gourmet Gaius Gavius Apicius. Malabathrum is among the spices that, according to Apicius, any good kitchen should contain. The famous Commagenum, an unguent produced in Commagene in present-day eastern Turkey, was made from goose-fat and aromatised with cinnamon oil and spikenard. Malobathrum from Egypt (Dioscorides I, 63) was based on cattle-fat and contained cinnamon as well; one pound cost 300 *denarii*. The Roman poet Martial (VI, 55) made fun of Romans who drip unguents, smell of cassia and cinnamon taken from a bird's nest, and look down on him who does not smell at all.

Before the foundation of Cairo, Alexandria was the Mediterranean shipping port of cinnamon. Europeans who knew the Latin writers who were quoting Herodotus knew cinnamon came up the Red Sea to the trading ports of Egypt, but whether from Ethiopia or not was less than clear. When the Sieur de Joinville accompanied his king to Egypt on crusade in 1248, he reported what he had been told—and believed—that cinnamon was fished up in nets at the source of the Nile out at the edge of the world. Through the Middle Ages, the source of cinnamon was a mystery to the Western world. Marco Polo avoided precision on this score. In Herodotus and other authors, Arabia was the source of cinnamon: giant cinnamon birds collected the cinnamon sticks from an unknown land where the cinnamon trees grew and used them to construct their nests; the Arabs employed a trick to obtain the sticks. This story was current as late as 1310 in Byzantium, although in the first century, Pliny the Elder had written that the traders had made this up in order to charge more. The first mention of the spice growing in Sri Lanka was in Zakariya al-Qazwini's *Athar al-bilad wa-akhbar al-'ibad* ("Monument of Places and History of God's Bondsmen") in about 1270. This was followed shortly thereafter by John of Montecorvino, in a letter of about 1292.

Indonesian rafts transported cinnamon (known in Indonesia as *kayu manis*- literally "sweet wood") on a "cinnamon route" directly from the Moluccas to East Africa, where local traders then carried it north to the Roman market.

Arab traders brought the spice by overland trade routes to Alexandria in Egypt, where it was bought by Venetian traders from Italy who held a monopoly on the spice trade in Europe. The disruption of this trade by the rise of other Mediterranean powers, such as the Mamluk sultans and the Ottoman Empire, was one of many factors that led Europeans to search more widely for other routes to Asia.

Portuguese traders finally landed in Ceylon (Sri Lanka) at the beginning of the 16th century and restructured the traditional production and management of cinnamon by the Sinhalese, who later held the monopoly for cinnamon in Ceylon. The Portuguese established a fort on the island in 1518 and protected their own monopoly for over a hundred years.

Dutch traders finally dislodged the Portuguese by allying with the inland Kingdom of Kandy. They established a trading post in 1638, took control of the factories by 1640, and expelled all remaining Portuguese by 1658. "The shores of the island are full of it", a Dutch captain reported, "and it is the best in all the Orient: when one is downwind of the island, one can still smell cinnamon eight leagues out to sea." (Braudel 1984, p. 215)

The Dutch East India Company continued to overhaul the methods of harvesting in the wild and eventually began to cultivate its own trees.

In 1767, Lord Brown of East India Company established Anjarakkandy Cinnamon Estate near Anjarakkandy in Cannanore (now Kannur) district of Kerala, and this estate became Asia's largest cinnamon estate.

The British took control of the island from the Dutch in 1796. However, the importance of the monopoly of Ceylon was already declining, as cultivation of the cinnamon tree spread to other areas, the more common cassia bark became more acceptable to consumers, and coffee, tea, sugar, and chocolate began to outstrip the popularity of traditional spices.

Cultivation

Figure: *Leaves from a wild cinnamon tree*

Global annual production of cinnamon and cassia amounts to 27,500-35,000 tons. *Cinnamomum verum* accounts for 7,500-10,000 tons of production, with the remainder produced by other species. Sri Lanka produces 80-90% of the world's supply of *C. verum*, but that is the only species grown there; *C. verum* is also cultivated on a commercial scale in Seychelles and Madagascar. Global production of the other species averages 20,000-25,000 tons, of which Indonesia produces around two-thirds of the total, with significant production in China. India and Vietnam are also minor producers.

Cinnamon is harvested by growing the tree for two years, then coppicing it. The following year, about a dozen shoots will form from the roots.

The branches harvested this way are processed by scraping off the outer bark, then beating the branch evenly with a hammer to loosen the inner bark. The inner bark is then pried out in long rolls. Only 0.5 mm (0.020 in) of the inner bark is used; the outer, woody portion is discarded, leaving metre-long cinnamon strips that curl into rolls ("quills") on drying. Once dry, the bark is cut into 5- to 10-cm (2- to 4-in) lengths for sale.

The bark must be processed immediately after harvesting while still wet. Once processed, the bark will dry completely in four to six hours, provided it is in a well-ventilated and relatively warm environment. A less than ideal drying environment encourages the proliferation of pests in the bark, which may then require treatment by fumigation. Bark treated this way is not considered to be of the same premium quality as untreated bark.

Sri Lanka cinnamon has a very thin, smooth bark with a light-yellowish brown colour and a highly fragrant aroma. In recent years in Sri Lanka, mechanical devices have been developed to ensure premium quality and worker safety and health, following considerable research by the universities in that country, led by the University of Ruhuna.

Grading

The Sri Lankan grading system divides the cinnamon quills into four groups:

- *Alba*, less than 6 mm (0.24 in) in diametre
- *Continental*, less than 16 mm (0.63 in) in diametre
- *Mexican*, less than 19 mm (0.75 in) in diametre
- *Hamburg*, less than 32 mm (1.3 in) in diametre

These groups are further divided into specific grades. For example, Mexican is divided into M00 000 special, M000000, and M0000, depending on quill diametre and number of quills per kg.

Any pieces of bark less than 106 mm (4.2 in) long are categorized as quillings. Featherings are the inner bark of twigs and twisted shoots.

Chips are trimmings of quills, outer and inner bark that cannot be separated, or the bark of small twigs.

Species

Figure: *Ceylon cinnamon (*Cinnamomum verum*) on the left, and Indonesian cinnamon (*Cinnamomum burmannii*) quills*

A number of species are often sold as cinnamon:

- *Cinnamomum verum* ("true cinnamon", Sri Lanka cinnamon or Ceylon cinnamon)
- *C. burmannii* (Korintje, Padang cassia, or Indonesian cinnamon)
- *C. loureiroi* (Saigon cinnamon, Vietnamese cassia, or Vietnamese cinnamon)
- *C. cassia* (Cassia or Chinese cinnamon)

Ceylon cinnamon, using only the thin inner bark, has a lighter brown colour, a finer, less dense, and more crumbly texture, and is considered to be more aromatic and more subtle in flavour than cassia. Cassia has a much stronger (somewhat harsher) flavour than Ceylon cinnamon, is generally a medium to light reddish brown, hard and woody in texture, and thicker (2–3 mm (0.079–0.118 in) thick), as all of the layers of bark are used.

Potentially toxic coumarin is present in much lower levels in *C. burmannii* due to its low essential oil content. Levels of coumarin in Ceylon cinnamon are much lower than those in cassia.

The barks, when whole, are easily distinguished, and their microscopic characteristics are also quite distinct. Ceylon cinnamon sticks (quills) have many thin layers and can easily be made into powder using a coffee or spice grinder, whereas cassia sticks are much harder. Indonesian cinnamon is often sold in neat quills made up of one thick layer, capable of damaging a spice or coffee grinder. Saigon cinnamon (*C. loureiroi*) and Chinese cinnamon (*C. cassia*) are always sold as broken pieces of thick bark, as the bark is not supple enough to be rolled into quills. The powdered bark is harder to distinguish, but if it is treated with tincture of iodine (a test for starch), little effect is visible with pure Ceylon cinnamon, but when Chinese cinnamon is present, a deep-blue tint is produced.

Flavour, Aroma and Taste

The flavour of cinnamon is due to an aromatic essential oil that makes up 0.5% to 1% of its composition. This essential oil is prepared by roughly pounding the bark, macerating it in sea water, and then quickly distilling the whole. It is of a golden-yellow colour, with the characteristic odour of cinnamon and a very hot aromatic taste. The pungent taste and scent come from cinnamic aldehyde or cinnamaldehyde (about 90% of the essential oil from the bark) and, by reaction with oxygen as it ages, it darkens in colour and forms resinous compounds. Other chemical components of the essential oil include ethyl cinnamate, eugenol (found mostly in the leaves), beta-caryophyllene, linalool, and methyl chavicol.

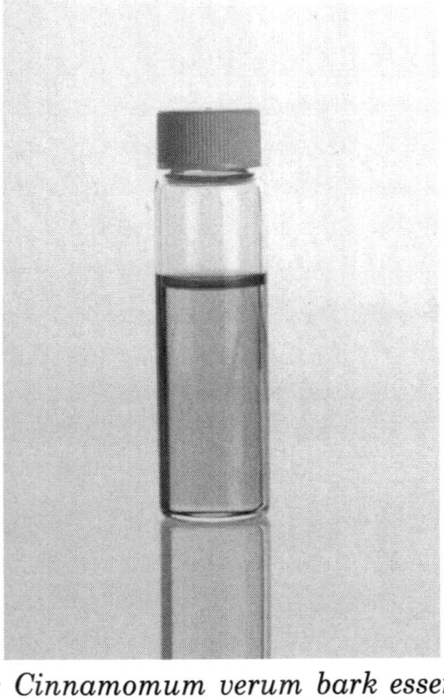

***Figure:** Cinnamomum verum bark essential oil*

Uses

Besides use as flavourant and spice in foods, cinnamon-flavoured tea, also flavoured with cardamom, is consumed as a hot beverage in Bangladesh, India and Pakistan.

Cinnamon bark is widely used as a spice. It is principally employed in cookery as a condiment and flavouring material. It is used in the preparation of chocolate, especially in Mexico, which is the main importer of cinnamon. It is also used in many dessert recipes, such as apple pie, doughnuts, and cinnamon buns as well as spicy candies, coffee, tea, hot cocoa, and liqueurs. True cinnamon, rather than cassia, is more suitable for use in sweet dishes.

In the Middle East, cinnamon is often used in savoury dishes of chicken and lamb. In the United States, cinnamon and sugar are often used to flavour cereals, bread-based dishes, such as toast, and fruits, especially apples; a cinnamon-sugar mixture is even sold separately for such purposes. It is also used in Turkish cuisine for both sweet and savoury dishes.

Cinnamon can also be used in pickling. Cinnamon bark is one of the few spices that can be consumed directly. Cinnamon powder has long been an important spice in Persian cuisine, used in a variety of thick soups, drinks, and sweets. It is often mixed with rosewater or other spices to make a cinnamon-based curry powder for stews or just sprinkled on sweet treats (most notably *Shole-zard*, Persian Ôáå ÒÑÏ). It is also used in sambar powder or BisiBelebath powder in Karnataka, which gives it a rich aroma and unique taste.

Cinnamon has been proposed for use as an insect repellent, although it remains untested. Cinnamon leaf oil has been found to be very effective in killing mosquito larvae. Of the compounds found in the essential oil from cinnamon leaves, cinnamyl acetate, eugenol, and anethole, and in particular cinnamaldehyde, were found to have the highest effectiveness against mosquito larvae.

Cinnamon, as a warm and dry substance, was believed by doctors in ancient times to cure snakebites, freckles, the common cold, and kidney troubles, among other ailments. However, newer studies showed that some substances in cinnamon, particularly coumarin, may cause liver damage in some sensitive people.

Cinnamon along with garlic is used as a fish and meat preservative and in the future might be used in an inner layer of plastic as it has antimicrobial properties up to 120 degree Celsius; they can also be used to preserve fried and deep fried foods.

Figure: *Cinnamon bark*

Use as an Alcohol Flavorant

It is a popular flavoring in numerous alcoholic beverages. Fireball Cinnamon Whisky is a cinnamon-flavoured whisky-based liqueur produced by the Sazerac Company. Somewhat similar products — Red Stag Spiced by Jim Beam, DeKuyper Hot Damn!, Gold Strike cinnamon liqueur containing gold snippets, produced by Lucas Bols, Hood River Distillers *SinFire Cinnamon Whisky*. and Goldschlager Cinnamon Schnapps containing gold flecks — which also have cinnamon as an ingredient in high potency liqueurs.

Other products feature cinnamon-infused vodka, including Smirnoff's Cinna-Sugar Twist. In November 2013, Beam's Pinnacle Vodka and Cinnabon teamed up to introduce their own brand of cinnamon flavored vodka, Cinnabon Vodka. Yet another is Stolichnaya Zinamom Vodka.

Cinnamon brandy concoctions, called "Cinnamon liqueur" and made with distilled alcohol, are popular in parts of Greece. For those who wish to make their own "cinnamon liqueur" there is much controversy concerning the proper ingredients. Particularly, some purported "cinnamon" (that found in "cinnamon sticks") isn't in fact cinnamon. as the latter may be banned or limited in some countries due to the presence of coumarin.

There are many cinnamon-infused liquors on the market.

Beneficial Effects

Cinnamon is used in traditional medicine, and several studies have tested chemicals extracted from cinnamon for various possible medicinal effects.

Anti-viral

In an experiment testing the effects of various plants used in traditional Indian medicine, an extract of *Cinnamomum cassia* had an effect on HIV-1. Another study found that eugenol, a chemical found in cinnamon essential oils, and in other plants, inhibited the replication of the virus causing herpes *in vitro*. The compound cinnzeylanine, from *C. zeylanicum*, also had antiviral properties in a model system using silkworm cells.

Diabetes

Two studies have shown that including cinnamon and cinnamon extract in the diet may help type 2 diabetics to control blood glucose levels. One study used *C. cassia*, while the other study used an extract (made from "Chinese *Cinnamomum aromaticum*", an older name for *C. cassia*). Apart from the most common flavanol (epi)catechin and (epi)afzelechin, cinnamon proanthocyanidins contain (epi)catechingallate and (epi)gallocatechin units. Furthermore, these proanthocyanidins are bioavailable and may have an effect on the target tissues. However a Cochrane review study published in 2012 found that cinnamon was not more effective than placebo in reducing glucose levels and glycosylated haemoglobin A1c (HbA1c) (a long-term measurement of glucose control in diabetes). Authors concluded that "There is insufficient evidence to support the use of cinnamon for type 1 or type 2 diabetes mellitus".

Antioxidant

Pharmacological experiments suggest that dietary cinnamon-derived cinnamic aldehyde (cinnamaldehyde) activates the Nrf2-dependent antioxidant response in human epithelial colon cells and may therefore represent an experimental chemopreventive dietary factor targeting colorectal carcinogenesis. Recent research documents antimelanoma activity of cinnamic aldehyde observed in cell culture and a mouse model of human melanoma.

Alzheimer's Disease

A 2011 study isolated a substance (CEppt) in the cinnamon plant that inhibits development of Alzheimer's disease in mice. CEppt, an extract of cinnamon bark, seems to treat a mouse model of Alzheimer's disease.

Adverse Effects

The European Food Safety Authority in 2008 considered toxicity of coumarin, known to cause liver and kidney damage in high concentrations and a significant component of cinnamon, and metabolic effect on humans with CYP2A6 polymorphism, and confirmed a maximum recommended Tolerable Daily Intake (TDI) of 0.1 mg of coumarin per kg of body weight. The European Union set a guideline for maximum coumarin content in foodstuffs of 50 mg per kg of dough in seasonal foods, and 15 mg per kg in everyday baked foods. These limits are low enough to affect the flavour of cinnamon pastries.

Complexion Problems

Over the ages, cinnamon has been used in unconventional medicine as a good way to treat skin problems. Cinnamon enjoys the largest popularity in Asia – especially in India and China. The minerals contained in this old spice sooth the skin and help to eliminate acne and pimples. Therefore, cinnamon is used as an ingredient in plenty of traditional beauty masks.

Nutritional Information

Ten grams (about 2.1 teaspoons) of ground cinnamon contain:

- Energy: 103.4 kJ (24.7 kcal)
- Fat: 0.12 g
- Carbohydrates: 8.06 g (of which - fibres: 5.31 g, sugars: 0.2 g)
- Protein: 0.4 g

Cloves

Cloves are the aromatic flower buds of a tree in the family Myrtaceae, *Syzygium aromaticum*. They are native to the Maluku Islands in Indonesia, and are commonly used as a spice. Cloves are commercially harvested primarily in Indonesia, India, Madagascar, Zanzibar, Pakistan, Sri Lanka—and the largest producer, Pemba Island, just off the coast of Tanzania.

The clove tree is an evergreen that grows up to 8–12 m tall, with large leaves and sanguine flowers grouped in terminal clusters. The flower buds initially have a pale hue, gradually turn green, then transition to a bright red when ready for harvest. Cloves are harvested at 1.5–2.0 cm long, and consist of a long calyx that terminates in four spreading sepals, and four unopened petals that form a small central ball.

Uses

Cloves are used in the cuisine of Asian, African, and the Near and Middle East, lending flavour to meats, curries, and marinades, as well as complement to fruit such as apples, pears, or rhubarb.

In Mexican cuisine, cloves are best known as *clavos de olor*, and often accompany cumin and cinnamon. About 85% of cloves' powerful taste is imparted by the chemical eugenol, and the quantity of the spice required is typically relatively small. It pairs well with cinnamon, allspice, vanilla, red wine, and basil, as well as onion, citrus peel, star anise, or peppercorns.

Nonculinary Uses

The spice is used in a type of cigarette called *kretek* in Indonesia. They have been smoked throughout Europe, Asia, and the United States. In 2009, clove cigarettes (as well as fruit- and candy-flavored cigarettes) were outlawed in the US. Cigarettes containing clove are now classified as cigars when sold in the US.

Clove may be used as an ant repellant.

They can be used as to make a fragrant pomander when combined with an orange.

Traditional Medicinal Uses

Cloves are used in Indian Ayurvedic medicine, Chinese medicine, and western herbalism and dentistry where the essential oil is used as an anodyne (painkiller) for dental emergencies. Cloves are used as a carminative, to increase hydrochloric acid in the stomach and to improve peristalsis. Cloves are also said to be a natural anthelmintic. The essential oil is used in aromatherapy when stimulation and warming are needed, especially for digestive problems. Topical application over the stomach or abdomen are said to warm the digestive tract. Applied to a cavity in a decayed tooth, it also relieves toothache.

In Chinese medicine, cloves or *ding xiang* are considered acrid, warm, and aromatic, entering the kidney, spleen and stomach meridians, and are notable in their ability to warm the middle, direct stomach *qi* downward, to treat hiccough and to fortify the kidney *yang*. Because the herb is so warming, it is contraindicated in any persons with fire symptoms and according to classical sources should not be used for anything except cold from *yang* deficiency. As such, it is used in formulas for impotence or clear vaginal discharge from *yang* deficiency, for morning sickness together with ginseng and patchouli, or for vomiting and diarrhea due to spleen and stomach coldness.

Cloves may be used internally as a tea and topically as an oil for hypotonic muscles, including for multiple sclerosis. This is also found in Tibetan medicine. Some recommend avoiding more than occasional use of cloves internally in the presence of *pitta* inflammation such as is found in acute flares of autoimmune diseases.

Modern Medicinal uses and Pharmaceutical Preparations

While it has been used historically in the West for dental pain, the U.S. Food and Drug Administration (FDA) has reclassified eugenol (one of the chemicals contained in clove oil), downgrading its effectiveness rating. The FDA now believes not enough evidence indicates eugenol is effective for toothache pain. Clove oil seems safe when applied to the skin.

Studies to determine its effectiveness for fever reduction, as a mosquito repellent, and to prevent premature ejaculation have been inconclusive. Clove may reduce blood sugar levels.

Tellimagrandin II is an ellagitannin found in *S. aromaticum* with antiherpesvirus properties.

The buds have antioxidant properties.

Clove oil can be used to anesthetize fish, and prolonged exposure to higher doses (the recommended dose is 400 mg/l) is considered a humane means of euthanasia.

In addition, clove oil is used in preparation of some toothpastes, laxative pills, and Clovacaine solution, which is a local anesthetic used in oral ulceration and inflammation. Eugenol (or clove oil generally) is mixed with zinc oxide to form a temporary tooth cavity filling.

Adulteration

Clove stalks are slender stems of the inflorescence axis that show opposite decussate branching. Externally, they are brownish, rough, and irregularly wrinkled longitudinally with short fracture and dry, woody texture.

Mother cloves (anthophylli) are the ripe fruits of cloves that are ovoid, brown berries, unilocular and one-seeded. This can be detected by the presence of much starch in the seeds. Brown cloves are expanded flowers from which both corollae and stamens have been detached.

Exhausted cloves have most or all the oil removed by distillation. They yield no oil and are darker in colour.

History

Archeologists have found cloves in a ceramic vessel in Syria, with evidence that dates the find to within a few years of 1721 BC. In the third century BC, a Chinese leader in the Han Dynasty required those who addressed him to chew cloves to freshen their breath. Cloves were traded by Muslim sailors and merchants during the Middle Ages in the profitable Indian Ocean trade, the clove trade is also mentioned by Ibn

Battuta and even famous *Arabian Nights* characters such Sinbad the Sailor are known to have bought and sold cloves from India.

Until modern times, cloves grew only on a few islands in the Maluku Islands (historically called the Spice Islands), including Bacan, Makian, Moti, Ternate, and Tidore. In fact, the clove tree that experts believe is the oldest in the world, named *Afo*, is on Ternate. The tree is between 350 and 400 years old. Tourists are told that seedlings from this very tree were stolen by a Frenchman named Poivre in 1770, transferred to France, and then later to Zanzibar, which is today the world's largest producer of cloves.

Until cloves were grown outside of the Maluku Islands, they were traded like oil, with a forced limit on exportation. As the Dutch East India Company consolidated its control of the spice trade in the 17th century, they sought to gain a monopoly in cloves as they had in nutmeg. However, "unlike nutmeg and mace, which were limited to the minute Bandas, clove trees grew all over the Moluccas, and the trade in cloves was way beyond the limited policing powers of the corporation."

Active Compounds

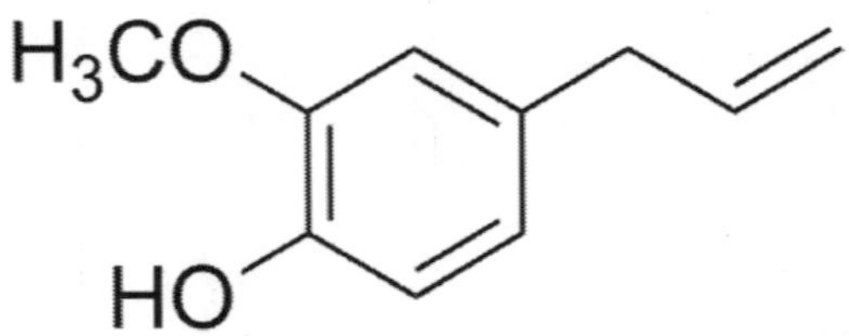

Figure: *The compound eugenol is responsible for most of the characteristic aroma of cloves.*

Eugenol comprises 72-90% of the essential oil extracted from cloves, and is the compound most responsible for the cloves' aroma. Other important essential oil constituents of clove oil include acetyl eugenol, beta-caryophyllene and vanillin, crategolic acid, tannins such as bicornin, gallotannic acid, methyl salicylate (painkiller), the flavonoids eugenin, kaempferol, rhamnetin, and eugenitin, triterpenoids such as oleanolic acid, stigmasterol, and campesterol, and several sesquiterpenes.

Eugenol can be toxic in relatively small quantities; with a dose of 5 - 10 ml severely affecting a 2 year old child.

Saffron

Saffron is a spice derived from the flower of *Crocus sativus*, commonly known as the saffron crocus. *Crocus* is a genus in the family Iridaceae. Saffron crocus grows to 20–30 cm (8–12 in) and bears up to four flowers, each with three vivid crimson stigmas, which are the distal end of a carpel. Together with the styles, or stalks that connect the stigmas to their host plant, the dried stigmas are used mainly in various cuisines as a seasoning and colouring agent.

Saffron, long among the world's most costly spices by weight, is native to Greece or Southwest Asia and was first cultivated in Greece. As a genetically monomorphic clone, it was slowly propagated throughout much of Eurasia and was later brought to parts of North Africa, North America, and Oceania.

The saffron crocus, unknown in the wild, probably descends from *Crocus cartwrightianus*, which originated in Crete; *C. thomasii* and *C. pallasii* are other possible precursors. The saffron crocus is a triploid that is "self-incompatible" and male sterile; it undergoes aberrant meiosis and is hence incapable of independent sexual reproduction—all propagation is by vegetative multiplication via manual "divide-and-set" of a starter clone or by interspecific hybridisation. If *C. sativus* is a mutant form of *C. cartwrightianus*, then it may have emerged via plant breeding, which would have selected for elongated stigmas, in late Bronze Age Crete.

Saffron's taste and iodoform- or hay-like fragrance result from the chemicals picrocrocin and safranal. It also contains a carotenoid dye, crocin, which imparts a rich golde n-yellow hue to dishes and textiles. Its recorded history is attested in a 7th-century BC Assyrian botanical treatise compiled under Ashurbanipal, and it has been traded and used for over four millennia. Iran now accounts for approximately 90% of the world production of saffron.

Figure: *Saffron – valuable stigmas, or threads, are tediously plucked, piled, and dried.*

Description

The domesticated saffron crocus, *Crocus sativus*, is an autumn-flowering perennial plant unknown in the wild. Its progenitors are possibly the eastern Mediterranean autumn-flowering *Crocus cartwrightianus*, which is also known as "wild saffron" and originated in Greece. The saffron crocus probably resulted when *C. cartwrightianus* was subjected to extensive artificial selection by growers seeking longer stigmas. *C. thomasii* and *C. pallasii* are other possible sources.

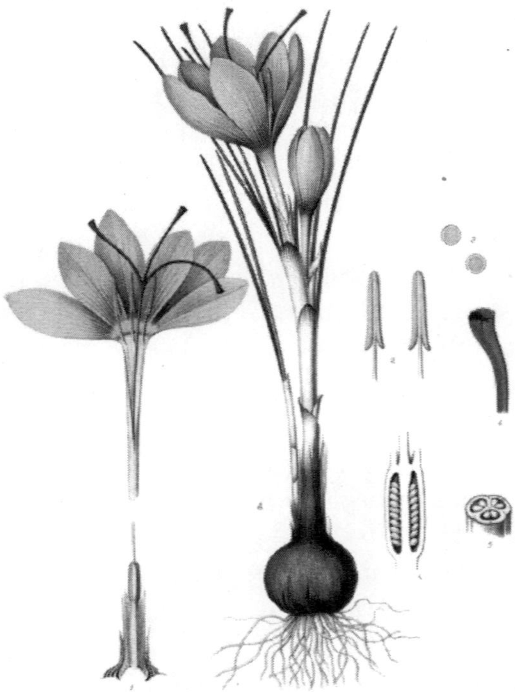

Figure: *Köhler's Medicinal Plants*

It is a sterile triploid form, which means that three homologous sets of chromosomes compose each specimen's genetic complement; *C. sativus* bears eight chromosomal bodies per set, making for 24 in total. Being sterile, the purple flowers of *C. sativus* fail to produce viable seeds; reproduction hinges on human assistance: clusters of corms, underground, bulb-like, starch-storing organs, must be dug up, divided, and replanted.

A corm survives for one season, producing via this vegetative division up to ten "cormlets" that can grow into new plants in the next season. The compact corms are small, brown globules that can measure as large as 5 cm (2.0 in) in diametre, have a flat base, and are shrouded in a dense mat of parallel fibres; this coat is referred to as the "corm tunic". Corms also bear vertical fibres, thin and net-like, that grow up to 5 cm above the plant's neck.

Figure: *C. sativus.*

The plant grows to a height of 20–30 cm (8–12 in), and sprouts 5–11 white and non-photosynthetic leaves known as cataphylls. These membrane-like structures cover and protect the crocus's 5 to 11 true leaves as they bud and develop. The latter are thin, straight, and blade-like green foliage leaves, which are 1–3 mm in diametre, either expand after the flowers have opened ("hysteranthous") or do so simultaneously with their blooming ("synanthous"). *C. sativus* cataphylls are suspected by some to manifest prior to blooming when the plant is irrigated relatively early in the growing season. Its floral axes, or flower-bearing structures, bear bracteoles, or specialised leaves that sprout from the flower stems; the latter are known as pedicels. After aestivating in spring, the plant sends up its true leaves, each up to 40 cm (16 in) in length. In autumn, purple buds appear. Only in October, after most other flowering plants have released their seeds, do its brilliantly hued flowers develop; they range from a light pastel shade of lilac to a darker and more striated mauve. The flowers possess a sweet, honey-like fragrance. Upon flowering, plants average less than 30 cm (12 in) in height. A three-pronged style emerges from each flower. Each prong terminates with a vivid crimson stigma 25–30 mm (0.98–1.18 in) in length.

Cultivation

Figure: *Saffron bulbs for vegetative reproduction*

Crocus sativus thrives in the Mediterranean maquis, an ecotype superficially resembling the North American chaparral, and similar climates where hot and dry summer breezes sweep semi-arid lands. It can nonetheless survive cold winters, tolerating frosts as low as "10 °C (14 °F) and short periods of snow cover. Irrigation is required if grown outside of moist environments such as Kashmir, where annual rainfall averages 1,000–1,500 mm (39–59 in); saffron-growing regions in Greece (500 mm or 20 in annually) and Spain (400 mm or 16 in) are far drier than the main cultivating Iranian regions. What makes this possible is the timing of the local wet seasons; generous spring rains and drier summers are optimal. Rain immediately preceding flowering boosts saffron yields; rainy or cold weather during flowering promotes disease and reduces yields. Persistently damp and hot conditions harm the crops, and rabbits, rats, and birds cause damage by digging up corms. Nematodes, leaf rusts, and corm rot pose other threats. Yet *Bacillus subtilis* inoculation may provide some benefit to growers by speeding corm growth and increasing stigma biomass yield.

Figure: *Saffron harvesting, Torbat-e Heydarieh, Iran*

The plants fare poorly in shady conditions; they grow best in full sunlight. Fields that slope towards the sunlight are optimal (i.e., south-sloping in the Northern Hemisphere). Planting is mostly done in June in the Northern Hemisphere, where corms are lodged 7–15 cm (2.8–5.9 in) deep; its roots, stems, and leaves can develop between October and February. Planting depth and corm spacing, in concert with climate, are critical factors in determining yields. Mother corms planted deeper yield higher-quality saffron, though form fewer flower buds and daughter corms. Italian growers optimise thread yield by planting 15 cm (5.9 in) deep and in rows 2–3 cm (0.79–1.18 in) apart; depths of 8–10 cm (3.1–3.9 in) optimise flower and corm production. Greek, Moroccan, and Spanish growers employ distinct depths and spacings that suit their locales.

C. sativus prefers friable, loose, low-density, well-watered, and well-drained clay-calcareous soils with high organic content. Traditional raised beds promote good drainage. Soil organic content was historically boosted via application of some 20–30 tonnes of manure per hectare. Afterwards, and with no further manure application, corms were planted. After a period of dormancy through the summer, the corms send up their narrow leaves and begin to bud in early autumn. Only in mid-autumn do they flower. Harvests are by necessity a speedy affair: after blossoming at dawn, flowers quickly wilt as the day passes. All plants bloom within a window of one or two weeks. Roughly 150 flowers together yield but 1 g (0.035 oz) of dry saffron threads; to produce 12 g

(0.42 oz) of dried saffron (or 72 g (2.5 oz) moist and freshly harvested), 1 kg (2.2 lb) of flowers are needed; 1 lb (0.45 kg) yields 0.2 oz (5.7 g) of dried saffron. One freshly picked flower yields an average 30 mg (0.0011 oz) of fresh saffron or 7 mg (0.00025 oz) dried.

Spice

Chemistry: Saffron contains more than 150 volatile and aroma-yielding compounds. It also has many nonvolatile active components, many of which are carotenoids, including zeaxanthin, lycopene, and various á- and â-carotenes. However, saffron's golden yellow-orange colour is primarily the result of á-crocin. This crocin is trans-crocetin di-(â-D-gentiobiosyl) ester; it bears the systematic (IUPAC) name 8,8-diapo-8,8-carotenoic acid. This means that the crocin underlying saffron's aroma is a digentiobiose ester of the carotenoid crocetin. Crocins themselves are a series of hydrophilic carotenoids that are either monoglycosyl or diglycosyl polyene esters of crocetin. Crocetin is a conjugated polyene dicarboxylic acid that is hydrophobic, and thus oil-soluble. When crocetin is esterified with two water-soluble gentiobioses, which are sugars, a product results that is itself water-soluble. The resultant á-crocin is a carotenoid pigment that may comprise more than 10% of dry saffron's mass. The two esterified gentiobioses make á-crocin ideal for colouring water-based and non-fatty foods such as rice dishes.

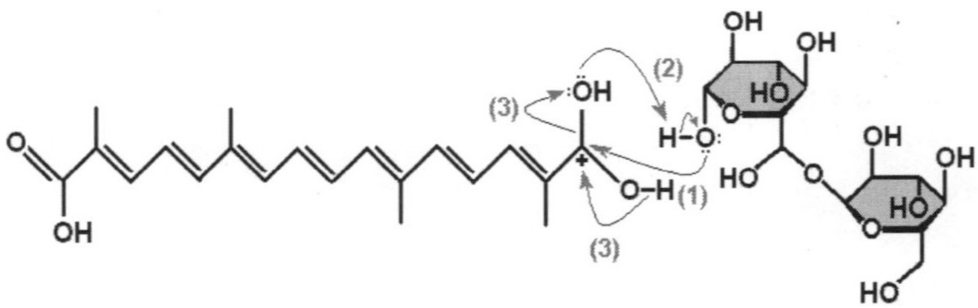

Figure: *Esterification reaction between crocetin and gentiobiose. Components of á–crocin*

The bitter glucoside picrocrocin is responsible for saffron's flavour. Picrocrocin (chemical formula: C16H26O7; systematic name: 4-(â-D-glucopyranosyloxy)-2,6,6- trimethylcyclohex-1-ene-1-carboxaldehyde) is a union of an aldehyde sub-element known as safranal (systematic name: 2,6,6-trimethylcyclohexa-1,3-diene-1-carboxaldehyde) and a carbohydrate. It has insecticidal and pesticidal properties, and may comprise up to 4% of dry saffron. Picrocrocin is a truncated version of the carotenoid zeaxanthin that is produced via oxidative cleavage, and is the glycoside of the terpene aldehyde safranal. The reddish-coloured zeaxanthin is, incidentally, one of the carotenoids naturally present within the retina of the human eye.

When saffron is dried after its harvest, the heat, combined with enzymatic action, splits picrocrocin to yield D–glucose and a free safranal molecule. Safranal, a volatile oil, gives saffron much of its distinctive aroma. Safranal is less bitter than picrocrocin and may comprise up to 70% of dry saffron's volatile fraction in some samples. A second element underlying saffron's aroma is 2-hydroxy-4,4,6-trimethyl-2,5-cyclohexadien-1-one, which produces a scent described as saffron, dried hay-like. Chemists find this is the most powerful contributor to saffron's fragrance, despite its presence in a lesser quantity than safranal. Dry saffron is highly sensitive to fluctuating pH levels, and rapidly breaks down chemically in the presence of light and oxidising agents. It must, therefore, be stored away in air-tight containers to minimise contact with atmospheric oxygen. Saffron is somewhat more resistant to heat.

Grades and ISO 3632 Categories

Figure: *Red threads and yellow styles.*

Saffron is not all of the same quality and strength. Strength is related to several factors including the amount of style picked along with the red stigma. Age of the saffron is also a factor. More style included means the saffron is less strong gram for gram, because the colour and flavour are concentrated in the red stigmas. Saffron from Iran, Spain and Kashmir is classified into various grades according to the relative amounts of red stigma and yellow styles it contains. Grades of Iranian saffron are: "sargol" (red stigma tips only, strongest grade), "pushal" or "pushali" (red stigmas plus some yellow style, lower strength), "bunch" saffron (red stigmas plus large amount of yellow style, presented in a tiny bundle like a miniature wheatsheaf) and "konge" (yellow style only, claimed to have aroma but with very little, if any, colouring potential). Grades of Spanish saffron are "coupé" (the strongest grade, like Iranian sargol), "mancha" (like Iranian pushal), and in order of further decreasing strength "rio", "standard" and "sierra" saffron. The word "mancha" in the Spanish classification can have two meanings: a general grade of saffron or a very high quality Spanish-grown saffron from a specific geographical origin. Real Spanish-grown La Mancha saffron has PDO protected status and this is displayed on the product packaging. Spanish growers fought hard for Protected Status because they felt that imports of Iranian saffron re-packaged in Spain and sold as "Spanish Mancha saffron" were undermining the genuine La Mancha brand.

Countries producing less saffron do not have specialised words for different grades and may only produce one grade. Artisan producers in Europe and New Zealand have offset their higher labour charges for saffron harvesting by targeting quality, only offering extremely high grade saffron.

In addition to descriptions based on how the saffron is picked, saffron may be categorised under the international standard ISO 3632 after laboratory measurement of crocin (responsible for saffron's colour), picrocrocin (taste), and safranal (fragrance or aroma) content. However, often there is no clear grading information on the product packaging and little of the saffron readily available in UK is labelled with ISO category. This lack of information makes it hard for customers to make informed choices when comparing prices and buying saffron.

Under ISO 3632, determination of non-stigma content ("floral waste content") and other extraneous matter such as inorganic material ("ash") are also key. Grading standards are set by the International Organization for Standardization, a federation of national standards bodies. ISO 3632 deals exclusively with saffron and establishes three categories: III (poorest quality), II, and I (finest quality). Formerly there was also category IV, which was below category III. Samples are assigned categories by gauging the spice's crocin and picrocrocin content, revealed by measurements of specific spectrophotometric absorbance. Safranal is treated slightly differently and rather than there being threshold levels for each category, samples must give a reading of 20-50 for all categories.

These data are measured through spectrophotometry reports at certified testing laboratories worldwide. Higher absorbances imply greater levels of crocin, picrocrocin and safranal, and thus a greater colouring potential and therefore strength per gram. The absorbance reading of crocin is known as the "colouring strength" of that saffron. Saffron's colouring strength can range from lower than 80 (for all category IV saffron) up to 200 or greater (for category I). The world's finest samples (the selected, most red-maroon, tips of stigmas picked from the finest flowers) receive colouring strengths in excess of 250, making such saffron over three times more powerful than category IV saffron. Market prices for saffron types follow directly from these ISO categories. Sargol and coupé saffron would typically fall into ISO 3632 category I. Pushal and mancha would probably be assigned to category II. On many saffron packaging labels, neither the ISO 3632 category nor the colouring strength (the measurement of crocin content) is displayed.

However, many growers, traders, and consumers reject such lab test numbers. Some people prefer a more holistic method of sampling batches of threads for taste, aroma, pliability, and other traits in a fashion similar to that practised by experienced wine tasters. However, ISO 3632 grade and colouring strength information allow consumers to make instant comparisons between the quality of different saffron brands, without needing to purchase and sample the saffron. In particular, consumers can work out value for money based on price per unit of colouring strength rather than price per gram, given the wide possible range of colouring strengths that different kinds of saffron can have.

Despite attempts at quality control and standardisation, an extensive history of saffron adulteration, particularly among the cheapest grades, continues into modern times. Adulteration was first documented in Europe's Middle Ages, when those found selling adulterated saffron were executed under the *Safranschou* code. Typical methods include mixing in extraneous substances like beets, pomegranate fibres, red-dyed silk fibres, or the saffron crocus's tasteless and odourless yellow stamens. Other methods included dousing saffron fibres with viscid substances like honey or vegetable oil to increase their weight. However, powdered saffron is more prone to adulteration, with turmeric, paprika, and other powders used as diluting fillers. Adulteration can also consist of selling mislabelled mixes of different saffron grades. Thus, in India, high-grade Kashmiri saffron is often sold and mixed with cheaper Iranian imports; these mixes are then marketed as pure Kashmiri saffron, a development that has cost Kashmiri growers much of their income.

Types

The various saffron crocus cultivars give rise to thread types that are often regionally distributed and characteristically distinct. Varieties (not varieties in the botanical sense) from Spain, including the tradenames "Spanish Superior" and "Creme", are generally mellower in colour, flavour, and aroma; they are graded by government-imposed standards. Italian varieties are slightly more potent than Spanish. The most intense varieties tend to be Iranian. Various "boutique" crops are available from New Zealand, France, Switzerland, England, the United States, and other countries—some of them organically grown. In the U.S., Pennsylvania Dutch saffron—known for its "earthy" notes—is marketed in small quantities.

Figure: *Saffron from different producer countries, picked and dried in different ways gives rise to different end qualities.*

Consumers may regard certain cultivars as "premium" quality. The "Aquila" saffron, or *zafferano dell'Aquila*, is defined by high safranal and crocin content, distinctive thread shape, unusually pungent aroma, and intense colour; it is grown exclusively on eight hectares in the Navelli Valley of Italy's Abruzzo region, near L'Aquila. It was first introduced to Italy by a Dominican monk from Inquisition-era Spain. But the biggest saffron cultivation in Italy is in San Gavino Monreale, Sardinia, where it is grown on 40 hectares, representing 60% of Italian production; it too has unusually high crocin, picrocrocin, and safranal content. Another is the "Mongra" or "Lacha" saffron of Kashmir (*Crocus sativus* 'Cashmirianus'), which is among the most difficult for consumers to obtain. Repeated droughts, blights, and crop failures in the Indian-controlled areas of Kashmir combine with an Indian export ban to contribute to its prohibitive overseas prices. Kashmiri saffron is recognisable by its dark maroon-purple hue; it is among the world's darkest, which hints at strong flavour, aroma, and colouring effect.

Trade

Almost all saffron grows in a belt bounded by the Mediterranean in the west, and the rugged region encompassing Iran and Kashmir in the east. The other continents, except Antarctica, produce smaller amounts. Some 300 t (300,000 kg) of dried whole threads and powder are gleaned yearly, of which 50 t (50,000 kg) is top-grade "coupe" saffron. Iran answers for around 90–93% of global production and exports much of it. A few of Iran's drier eastern and southeastern provinces, including Fars, Kerman, and those in the Khorasan region, glean the bulk of modern global production. In 2005, the second-ranked Greece produced 5.7 t (5,700.0 kg), while Morocco and Kashmir, tied for third rank, each produced 2.3 t (2,300.0 kg).

In recent years, Afghan cultivation has risen. Azerbaijan, Morocco, and Italy are, in decreasing order, lesser producers. Prohibitively high labour costs and abundant Iranian imports mean that only select locales continue the tedious harvest in Austria, England, Germany, and Switzerland—among them the Swiss village of Mund, whose annual output is a few kilograms. Tasmania, China, Egypt, England (at a tiny village in the county of Norfolk) France, Israel, Mexico, New Zealand, Turkey (mainly around the town of Safranbolu), California, and Central Africa are microscale cultivators.

To glean 1 lb (450 g) of dry saffron requires the harvest of 50,000–75,000 flowers; a kilogram requires 110,000–170,000 flowers. Forty hours of labour are needed to pick 150,000 flowers. Stigmas are dried quickly upon extraction and (preferably) sealed in airtight containers. Saffron prices at wholesale and retail rates range from US$500 to US$5,000 per pound, or US$1,100–11,000/kg, equivalent to £2,500/€3,500 per pound or £5,500/€7,500 per kilogram. The price in Canada recently rose to CA$18,000 per kilogram. In Western countries, the average retail price in 1974 was $1,000/£500/€700 per pound, or US$2,200/£1,100/€1,550 per kilogram. In February, 2013, a retail bottle containing 0.06 ounces could be purchased for $16.26 or the equivalent of $4,336 per pound or as little as about $2,000/pound in larger quantities. A pound contains between 70,000 and 200,000 threads. Vivid crimson colouring, slight moistness, elasticity, and lack of broken-off thread debris are all traits of fresh saffron.

Use

Saffron's aroma is often described by connoisseurs as reminiscent of metallic honey with grassy or hay-like notes, while its taste has also been noted as hay-like and sweet. Saffron also contributes a luminous yellow-orange colouring to foods. Saffron is widely used in Indian, Persian, European, Arab, and Turkish cuisines. Confectioneries and liquors also often include saffron. Common saffron substitutes include safflower (*Carthamus tinctorius*, which is often sold as "Portuguese saffron" or "açafrão"), annatto, and turmeric (*Curcuma longa*). Saffron has also been used as a fabric dye, particularly in China and India, and in perfumery. It is used for religious purposes in India, and is widely used in cooking in many cuisines, ranging from the Milanese *risotto* of Italy to the *bouillabaisse* of France to the *biryani* with various meat accompaniments in South Asia.

Saffron has a long medicinal history as part of traditional healing; several modern research studies have hinted that the spice has possible anticarcinogenic (cancer-suppressing), anti-mutagenic (mutation-preventing), immunomodulating, and antioxidant-like properties. Saffron stigmas, and even petals, may be helpful for depression. Early studies show that saffron may protect the eyes from the direct effects of bright light and retinal stress apart from slowing down macular degeneration and retinitis pigmentosa. (Most saffron-related research refers to the stigmas, but this is often not made explicit in research papers.) Other controlled research studies have indicated that saffron may have many potential medicinal properties.

Cumin (Jeera)

Cumin; sometimes spelled cummin; *Cuminum cyminum*) is a flowering plant in the family Apiaceae, native from the east Mediterranean to India. Its seeds (each one contained within a fruit, which is dried) are used in the cuisines of many different cultures, in both whole and ground form.

Cumin is the dried seed of the herb *Cuminum cyminum*, a member of the parsley family. The cumin plant grows to 30–50 cm (12–20 in) tall and is harvested by hand. It is an annual herbaceous plant, with a slender, glabrous, branched stem which is 20–30 cm (8–12 in) tall and has a diametre of 3–5 cm (1 ^{1}D $_{4}$–2 in). Each branch has two to three sub-branches. All the branches attain the same height, therefore the plant has a uniform canopy. The stem is coloured grey or dark green. The leaves are 5–10 cm (2–4 in) long, pinnate or bipinnate, with thread-like leaflets. The flowers are small, white or pink, and borne in umbels.

Each umbel has five to seven umbellts. The fruit is a lateral fusiform or ovoid achene 4–5 mm (D $_6$–^{1}D $_5$ in) long, containing two mericarps with a single seed. Cumin seeds have eight ridges with oil canals. They resemble caraway seeds, being oblong in shape, longitudinally ridged, and yellow-brown in colour, like other members of the umbelliferae family such as caraway, parsley and dill.

History

Figure: *Cumin seeds*

Cumin has been in use since ancient times. Seeds excavated at the Indian site have been dated to the second millennium BC. They have also been reported from several New Kingdom levels of ancient Egyptian archaeological sites. In the ancient Egyptian civilisation cumin was used as spice and as preservative in mummification.

Originally cultivated in Iran and the Mediterranean region, cumin is mentioned in the Bible in both the Old Testament (Isaiah 28:27) and the New Testament (Matthew 23:23). The ancient Greeks kept cumin at the dining table in its own container (much as pepper is frequently kept today), and this practice continues in Morocco. Cumin was also used heavily in ancient Roman cuisine. It was introduced to the Americas by Spanish and Portuguese colonists. There are several different types of cumin but the most famous ones are black and green cumin which are both used in Persian cuisine.

Today, the plant is mostly grown in China, Uzbekistan, Tajikistan, Iran, Turkey, Morocco, Egypt, Syria, Mexico, Chile and India. Since cumin is often used as part of birdseed and exported to many countries, the plant can occur as a rare casual in many territories including Britain. Cumin occurs as a rare casual in the British Isles, mainly in

Southern England; but the frequency of its occurrence has declined greatly. According to the Botanical Society of the British Isles' most recent Atlas, only one record has been confirmed since 2000.

In India, cumin has been used for millennia as a traditional ingredient of innumerable *kormas*, *masalas*, soups, and other spiced gravies.

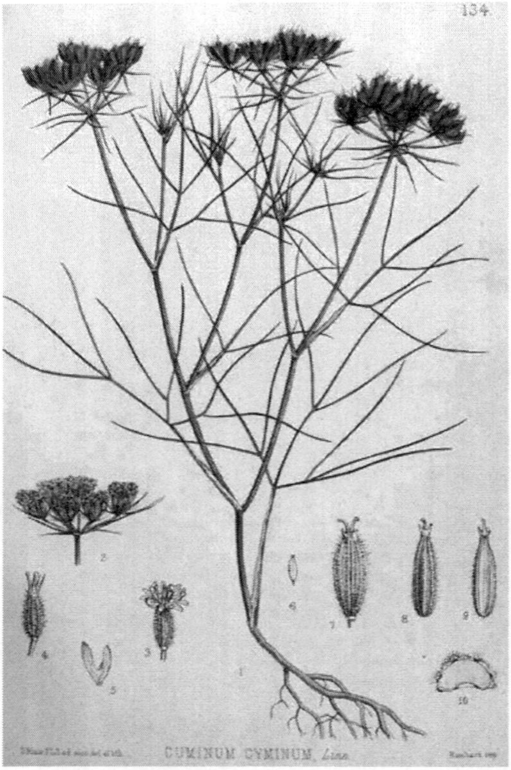

Figure: Cuminum cyminum *Linn.*

Cultivation and Production

Cultivation Areas: The main producer and consumer of cumin is India. It produces 70% of the world production and consumes 90% of its own production (which is 63% of the world production). Other producers are Syria (7%), Turkey (6%) and Iran (6%). The remaining 11% production is assigned to other countries. Totally, around 300,000 tons of cumin per year are produced worldwide. 2007 India produced around 175,000 tons of cumin on an area of about 410,000 ha. I.e. the average yield is 0.43 tons per hectare.

Climatic Requirements

Cumin is a drought tolerant, tropic or semi-tropic crop. Its origin is most probably Egypt, Turkmenistan and the east Mediterranean.

Cumin has a short growth season of 100 – 120 days. The optimum growth temperature ranges are between 25° and 30°C. The Mediterranean climate is most suitable for its growth; cumin requires a moderatly cool and dry climate. Cultivation of cumin requires a long, hot summer of three to four months. Upon low temperatures leaf colour changes from green to purple. High temperature might reduce growth period and induce early ripening. In India, Cumin is sown from October until the begin of December and harvesting starts in February. In Syria and Iran Cumin is sown from mid-November until mid-December (extensions up to mid-January are possible) and harvested in June/July.

Cultivation Parametres

Cumin is grown from seeds. The seeds need 2 to 5 °C (36 to 41 °F) for emergence, an optimum of 20–30 °C (68–86 °F) is suggested. Cumin is vulnerable to frost damage, especially at flowering and early seed formation stages. Methods to reduce frost damage are spraying with sulphuric acid (0.1%), irrigating the crop prior to frost incidence, setting up windbreaks or creating an early morning smoke cover. The seedlings of cumin are rather small and their vigor is low.

Soaking the seeds for 8 hours before sowing enhances germination. For an optimal plant population a sowing density of 12–15 kilograms per hectare (11–13 lb/acre) is recommended. Fertile, sandy, loamy soils with good aeration, proper drainage and high oxygen availability are preferred. The pH optimum of the soil ranges from pH 6.8 to 8.3. Cumin seedlings are sensitive to salinity and emergence from heavy soils is rather difficult for cumin. Therefore a proper seedbed preparation (smooth seed bed) is crucial for optimal establishment of cumin.

Two sowing methods are used for cumin, broadcasting and line sowing. For broadcast sowing, the field is divided into beds and the seeds are uniformly broadcast in this bed. Afterwards they are covered with soil using a rake.

For line sowing shallow furrows are prepared with hooks in a distance of 20 to 25 cm (8 to 10 in). The seeds are afterwards placed in these furrows and covered with soil. Line sowing offers advantages for intercultural operations such as weeding, hoeing or spraying.

The recommended sowing depth is 1–2 cm and the recommended sowing density is around 120 plants per square metre. The water requirements of cumin are lower than those of many other species. Despite, cumin is often irrigated after sowing to be sure that enough moisture is available for seedling development. The amount and frequency of irrigation depends on the climate conditions.

Cultivation Management

The relative humidity in the centre of origin of cumin is rather low. High relative humidity (i.e. wet years) favours fungal diseases. Cumin is especially sensitive to *Alternaria* blight and *Fusarium* wilt. Early sown crops are exhibit stronger disease effects than late sown crops. The most important disease is wilt caused by *Fusarium* resulting in yield losses up to 80%. *Fusarium* is seed- or soil-borne and it requires distinct soil temperatures for development of epidemics. Inadequate fertilization might favour *Fusarium* epidemics. Cumin blight (*Alternaria*) appears in the form of dark brown spots on leaves and stems. When the weather is cloudy after flowering the incidence of the disease is increased. Another, but less important disease is powdery mildew. Incidence of powdery mildew in early development can cause drastic yield losses because no seeds are formed. Later in development powdery mildew causes discoloured, small seeds.

Pathogens can lead to high reductions in crop yield. Cumin can be attacked by aphids (*Myzus persicae*) at flowering stage. They suck the sap of the plant from tender parts and flowers. The plant becomes yellow, the seed formation is reduced (yield reduction) and the quality of the harvested product decreases. Heavily infested plant parts should be removed. Other important pests are the mites (*Petrobia latens*) which frequently attack the crop. Since the mites mostly feed on young leaves, the infestation is more severe on young inflorescences.

The open canopy of cumin is another problem. Only a low proportion of the incoming light is absorbed. The Leaf Area Index (LAI) of cumin is low (approximately 1.5). This might be a problem because weeds can compete with cumin for essential resources such as water and light and thereby lower yield. The slow growth and a short stature of cumin favours weed competition additionally. Two hoeing and weeding sessions (30 and 60 days after sowing) are needed for the control of weeds. During the first weeding session (30 days after sowing) thinning should be done as well to remove excess plants. The use of pre-plant or pre-emergence herbicides is very effective in India. But this kind of herbicide application requires soil moisture for a successful weed control.

Fertilization recommendations in India

- 20 kg/ha (18 lb/acre) of phosphate (sowing)
- 30 kg/ha (27 lb/acre) of nitrogen, either

  single dose (30 days after sowing) or

  two doses (30 and 60 days after sowing)

Fertilization recommendations in Syria

- 50 kg (110 lb) of triple super phosphate (at planting)
- 50 kg of urea (at planting)

Breeding of Cumin

Cumin is a diploid species and with 14 chromosomes (i.e. 2n = 14). The chromosomes of the different varieties have morphological similarities and there is no distinct variation in length and volume. Most of the varieties available today are selections. The variabilities of yield and yield components are high. Varieties are developed by sib mating in enclosed chambers or by biotechnoloy. Cumin is a cross-pollinator, i.e. the breeds are already hybrids. Therefore, methods used for breeding are in vitro regenerations, DNA technologies and gene transfers. The in vitro cultivation of cumin allows the production of genetically identical plants. The main sources for the explants used in vitro regenerations are embryos, hypocotyl, shoot internodes, leaves and cotyledons. One goal of cumin breeding is to improve its resistance to biotic (fungal diseases) and abiotic (cold, drought, salinity) stresses. The potential genetic variability for conventional breeding of cumin is limited and research about cumin genetics is scarce.

Uses

Cumin seeds are used as a spice for their distinctive flavour and aroma. It is globally popular and an essential flavouring in many cuisines, particularly South Asian, Northern African and Latin American cuisines. Cumin can be found in some cheeses, such as Leyden cheese, and in some traditional breads from France. It is commonly used in traditional Brazilian cuisine. Cumin can be an ingredient in chili powder (often Tex-Mex or Mexican-style), and is found in *achiote* blends, *adobos, sofrito, garam masala*, curry powder, and *bahaarat*.

Cumin can be used ground or as whole seeds. It helps to add an earthy and warming feeling to food, making it a staple in certain stews and soups, as well as spiced gravies such as chili. It is also used as an ingredient in some pickles and pastries.

Medicinal Uses

In Sanskrit, Cumin is known as *Jira*. Jira means "that which helps digestion". In Ayurvedic system of medicine, dried Cumin seeds are used for medicinal purposes. The dried cumin seeds are powdered and used in different forms like *kashaya* (decoction), *arishta* (fermented decoction), *vati*(tablet/pills), and processed with ghee (a semi-fluid clarified butter). It is used internally and sometimes for external application also. It is known for its actions like enhancing appetite, taste perception, digestion, vision, strength, and lactation. It is used to treat diseases like fever, loss of appetite, diarrhea, vomiting, abdominal distension, edema and puerperal disorders.

A popular drink in southern India such as Kerala and Tamil Nadu is called jira water, made by boiling cumin seeds in water. It is understood that cumin is beneficial for heart disease, swellings, tastelessness, vomiting, poor digestion and chronic fever.

Ahmad Reza Gohari and Soodabeh Saeidnia have reviewed the Phytochemistry of Cuminum cyminum seeds and its standards. They have reported many pharmacological effects such as anti-diabetic, Immunologic, anti-epileptic, anti-tumor and antimicrobial activities.' A study by researchers at Mysore University in India reports the potential anti-diabetic properties of cumin.

Efraim Lev and Zohar Amar have reported several medicinal properties and health benefits of cumin seeds. According to the authors, cumin seeds and warm jeera water is believed to improve saliva secretion, provide relief in digestive disorders.

Secondary Metabolites

Cuminaldehyde, cymene and terpenoids are the major volatile components of cumin. Results of a study conducted in India showed that cumin can be used as an antioxidant. The antioxidative potential is correlated with the phenol content of cumin. Cuminaldehyde has also antimicrobial and antifungal properties which could be shown e.g. with *Escherichia coli* and *Penicillium chrysogenum*.

Although cumin seeds contain a relatively large percentage of iron, extremely large quantities of cumin would need to be consumed for it to serve as a significant dietary source.

According to the USDA, one tablespoon of cumin spice contains

- 22 kcal (92 kJ) of food energy,
- 1.34 g (0.047 oz) of fat,
- 2.63 g (0.093 oz) of carbohydrates,
- 0.6 g (0.021 oz) of fibre and
- 1.07 g (0.038 oz) of protein.

Confusion with Other Spices

Cumin is sometimes confused with caraway (*Carum carvi*), another umbelliferous spice. Cumin, though, is hotter to the taste, lighter in colour, and larger. Many European languages do not distinguish clearly between the two. Many Slavic and Uralic languages refer to cumin as "Roman caraway". Examples include Czech: *kmín* – caraway, *øímský kmín* -cumin; Polish: *kminek* – caraway, *kmin rzymski* – cumin; Hungarian: *köméný* – caraway, *római köméný* – cumin. Finnish: *kumina* – caraway, *roomankumina* – cumin, although sometimes also called *juustokumina*, cheese caraway. In Norwegian, caraway is called both *karve* and *kummin/kømming* while cumin is *spisskummen*, from the word *spise*, to eat. Similarly in Swedish and Danish, caraway is *kummin/kommen*, while cumin is *spiskummin/spidskommen*. In German, *Kümmel* stands for caraway and *Kreuzkümmel* denotes cumin. In Icelandic, caraway is *kúmen*, while cumin is *kúmín*. In Romanian, *chimen* is caraway, while *chimion* is cumin.

The distantly related *Bunium persicum* and the unrelated *Nigella sativa* are both sometimes called black cumin (*q.v.*).

***Figure:** Black cumin seeds*

Aroma Profile

Cumin's distinctive flavour and strong, warm aroma are due to its essential oil content. Its main constituent aroma compounds are cuminaldehyde (a promising agent against alpha-synuclein aggregation) and cuminic alcohol. Other important aroma compounds of toasted cumin are the substituted pyrazines, 2-ethoxy-3-isopropylpyrazine, 2-methoxy-3-*sec*-butylpyrazine, and 2-methoxy-3-methylpyrazine. Other components include ã-terpinene, safranal, p-cymene and â-pinene.

5

Roots and Rhizomes

The common spices which are derived from roots, bulbs and rhizomes are turmeric, ginger, onion, garlic, horseradish and wasabi. Turmeric, a perennial herb grown as an annual crop, is ready for digging when the lower leaves turn yellow. The length of time to harvest maturity is dependent on cultivar. Hand digging is the most common form of harvesting the rhizomes although diggers and lifters can be used. The leaves must be cut prior to mechanical lifting or after hand digging. Care is necessary to avoid damage to the rhizome (splitting or bruising) as injuries can result in fungal infection and rejection. Rhizomes are lifted whole, washed, sun dried and the rhizome fingers (which are kept as seed material) are separated from the mother rhizome.

The harvesting of ginger, a perennial herb but often grown as an annual, depends on the cultivar and varies from 7-9 months for annual crops while perennial crops are harvested at the growers' discretion. The time to harvest is also dependent on the projected end-use with fresh ginger with low fibre content harvested at about 190 days after planting. The harvesting method must ensure there is no rhizome damage and both manual and mechanical methods of rhizome digging are used. Some production areas use both methods by mechanically loosening the rhizomes to assist the hand lifting. Following lifting the rhizomes are washed, the roots removed and then killed by immersion in boiling water for about 10 minutes, dried and then stored.

The maturity of onions for harvest is based on top senescence when the bulb is converted from active growth to dormancy by under cutting and the beginning of the drying process. Garlic can be harvested at different stages of maturity for speciality markets but most commonly when the bulbs are well mature and the tops have dried completely. Horseradish, a perennial that is grown for its enlarged root, is ready for harvest after the leaves have been killed by frost. Horseradish that is harvested while it is still actively growing does not store as well.

Turmeric

Turmeric (*Curcuma longa*) is a rhizomatous herbaceous perennial plant of the ginger family, Zingiberaceae. It is native in southeast India, and needs temperatures between 20 °C and 30 °C (68 °F and 86 °F) and a considerable amount of annual rainfall to thrive. Plants are gathered annually for their rhizomes, and propagated from some of those rhizomes in the following season.

When not used fresh, the rhizomes are boiled for about 30–45 minutes and then dried in hot ovens, after which they are ground into a deep orange-yellow powder commonly used as a spice in Indian cuisine and even curries, for dyeing, and to impart colour to mustard condiments. One active ingredient in it is curcumin. It has a distinctly earthy, slightly bitter, slightly hot peppery flavour and a mustardy smell. Curcumin may treat: cancer, Alzheimer's disease, diabetes, allergies, arthritis and other chronic illnesses.

Figure: *Botanical view of* Curcuma longa.

India is a significant producer of turmeric which has regional names based on language and country.

As turmeric is a natural botanical product, it is not patentable.

Botanical Description

Appearance: Turmeric is a perennial herbaceous plant, which reaches a stature of up to 1 metre. There are highly branched, yellow to orange, cylindrical, aromatic rhizomes . The leaves are alternate and arranged in two rows . They are divided into leaf sheath, petiole and leaf blade. From the leaf sheaths, a false stem is formed. The petiole is 50 to 115 cm long. The simple leaf blades are usually of a length of 76 to 115 cm and rarely up to 230 cm. They have a width of 38 to 45 cm and are oblong to elliptic narrowing at the tip .

Inflorescence, Flower and Fruit

In China, the flowering time is usually in August. Terminally on the false stem there is a 12 to 20 cm long inflorescence stem containing many flowers. The bracts are light green and ovate with a length of 3 to 5 centimetres to oblong with a blunt upper end.

At the top of the inflorescence stem bracts are present on which there are no flowers, these are, white to green and sometimes tinged reddish-purple and its upper end is tapered.

The hermaphrodite flowers are zygomorphic and threefold. The three 0.8 to 1.2 centimetres long, sepals are fused, white, have fluffy hairs and the three calyx teeth are unequal. The three bright yellow petals are fused into a corolla tube up to 3 centimetres long. The three corolla lobes have a length of 1 to 1.5 cm, triangular with soft-spiny upper end. While the average corolla lobe is larger than the two lateral. Only the median stamen of the inner circle is fertile. The dust bag is spurred at its base. All other stamens are converted to staminodes. The outer staminodes are shorter than the labellum. The labellum is yellowish, with a yellow ribbon in its centre and it is obovate, with a length from 1.2 to 2 cm. Three carpels are under a constant, trilobed ovary adherent, which is sparsely hairy.

The fruit capsule opens with three compartments.

Biochemical Composition

The most important chemical components of turmeric are a group of compounds called curcuminoids, which include curcumin (diferuloylmethane), demethoxycurcumin, and bisdemethoxycurcumin. The best studied compound is curcumin, which constitutes 3.14% (on average) of powdered turmeric. In addition there are other important volatile oils such as turmerone, atlantone, and zingiberene. Some general constituents are sugars, proteins, and resins.

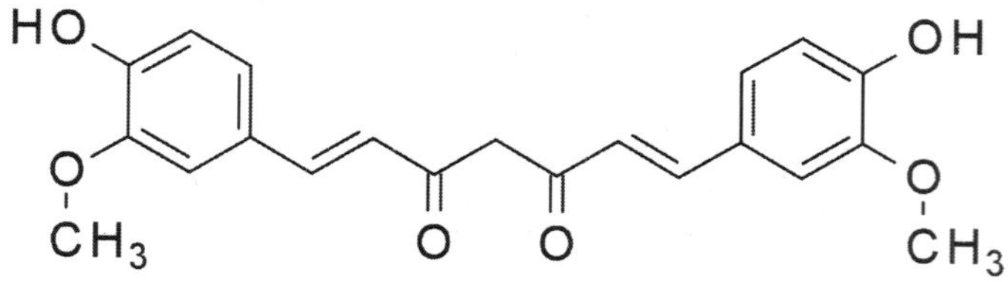

Figure: *Curcumin keto form*

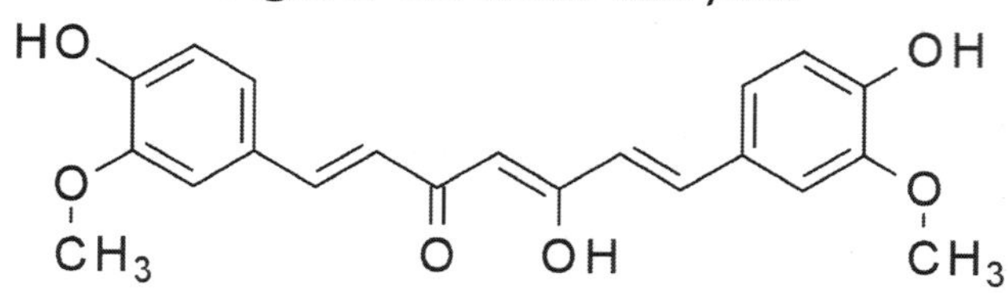

Figure: *Curcumin enol form*

Uses

Culinary: Turmeric grows wild in the forests of South and Southeast Asia. It is one of the key ingredients in many Asian dishes. Tamil traditional medicine, called Siddha, has recommended turmeric for medicine. Its use as a colouring agent is not of primary value in South Asian cuisine.

Turmeric is mostly used in savory dishes, but is used in some sweet dishes, such as the cake *Sfouf*. In India, turmeric plant leaf is used to prepare special sweet dishes, *patoleo*, by layering rice flour and coconut-jaggery mixture on the leaf, and then closing and steaming it in a special copper steamer (goa).

In recipes outside South Asia, turmeric is sometimes used as an agent to impart a rich, custard-like yellow colour. It is used in canned beverages and baked products, dairy products, ice cream, yogurt, yellow cakes, orange juice, biscuits, popcorn colour, sweets, cake icings, cereals, sauces, gelatins, etc. It is a significant ingredient in most commercial curry powders.

Most turmeric that is used is in the form of rhizome powder, in some regions (especially in Maharashtra, Goa, Konkan and Kanara), turmeric leaves are used to wrap and cook food. This use of turmeric leaves usually takes place in areas where turmeric is grown locally, since the leaves used are freshly picked. Turmeric leaves impart a distinctive flavour.

Although typically used in its dried, powdered form, turmeric is also used fresh, like ginger. It has numerous uses in Far Eastern recipes, such as pickle made from fresh turmeric that contains large chunks of soft turmeric.

Turmeric is widely used as a spice in South Asian and Middle Eastern cooking. Many Persian dishes use turmeric as a starter ingredient. Almost all Iranian fried dishes consist of oil, onions, and turmeric followed by any other ingredients that are to be included.

In Nepal, turmeric is widely grown and extensively used in many vegetable and meat dishes for its colour as well as for its potential value in traditional medicine.

In South Africa, turmeric is used to give boiled white rice a golden colour.

In Vietnam, turmeric powder is used to colour, and enhance the flavors of, certain dishes, such as bánh xèo, bánh khÍt and mi quang. The powder is also used in many other Vietnamese stir fried and soup dishes.

In Indonesia, the turmeric leaves are used for Minangese or Padangese curry base of Sumatra, such as *rendang, sate padang* and many other varieties.

In Thailand, fresh turmeric rhizomes are widely used in many dishes, in particular in the southern Thai cuisine, such as the yellow curry and turmeric soup.

In medieval Europe, turmeric became known as Indian saffron because it was widely used as an alternative to the far more expensive saffron spice.

Folk Medicine and Traditional Uses

In Tamil Nadu, turmeric has been used traditionally for thousands of years as a remedy for stomach and liver ailments, as well as topically to heal sores, basically for its supposed antimicrobial property. In the Siddha system (since c. 1900 BCE) turmeric was a medicine for a range of diseases and conditions, including those of the skin, pulmonary, and gastrointestinal systems, aches, pains, wounds, sprains, and liver disorders. A fresh juice is commonly used in many skin conditions, including eczema, chicken pox, shingles, allergy, and scabies.

Manjal Pal (turmeric milk) is warm milk mixed with some turmeric powder. It is commonly used in Tamil Nadu as a home remedy when someone is suffering from fever. Turmeric paste is often used in Tamil Nadu as an antiseptic in open wounds, while *chun-holud* (turmeric with slaked lime) is used to stop bleeding as home remedies. It is also used as a detanning agent in Tamil Nadu.

The active compound curcumin is believed to have a wide range of biological effects including anti-inflammatory, antioxidant, antitumour, antibacterial, and antiviral activities, which indicate potential in clinical medicine. In Chinese medicine, it is used for treatment of various infections and as an antiseptic.

Preliminary Medical Research

According to the National Centre for Complementary and Alternative Medicine, "there is little reliable evidence to support the use of turmeric for any health condition because few clinical trials have been conducted."

Although trials are ongoing for the use of turmeric to treat cancer, doses needed for any effect are difficult to establish in humans. As of December 2013, turmeric is being evaluated for its potential efficacy against several human diseases in clinical trials, including kidney and cardiovascular diseases, arthritis, several types of cancer and irritable bowel disease.

Specifically, turmeric is also being investigated in relation to Alzheimer's disease, diabetes, and other clinical disorders.

However, according to various basic research studies, administration of curcumin or turmeric can suppress several stages of cancer development in multiple tumor models. One study of curcumin on human cancer cells in vitro used hybrid molecules with the anti-nausea drug thalidomide to induce apoptosis in myeloma cancer cells. Some research shows compounds in turmeric to have anti-fungal and antibacterial properties; however, curcumin is not one of them.

Curcumin, the active component of turmeric, has also been shown to be a vitamin D receptor ligand "with implications for colon cancer chemoprevention."

Dye

Turmeric makes a poor fabric dye, as it is not very light fast. However, turmeric is commonly used in Indian and Bangladeshi clothing, such as saris and Buddhist monks' robes. Turmeric (coded as E100 when used as a food additive) is used to protect food products from sunlight. The oleoresin is used for oil-containing products. A curcumin and polysorbate solution or curcumin powder dissolved in alcohol is used for water-containing products. Over-colouring, such as in pickles, relishes, and mustard, is sometimes used to compensate for fading.

In combination with annatto (E160b), turmeric has been used to colour cheeses, yogurt, dry mixes, salad dressings, winter butter and margarine. Turmeric is also used to give a yellow colour to some prepared mustards, canned chicken broths and other foods (often as a much cheaper replacement for saffron).

Ceremonial Uses

Turmeric is considered highly auspicious and holy in India and has been used extensively in various Hindu ceremonies for centuries. Even today it is used in every part of India during wedding ceremonies and religious ceremonies.

Turmeric has played an important role in Hindu spiritualism. The robes of the Hindu monks were traditionally coloured with a yellow dye made of turmeric. Because of its yellow-orange colouring, turmeric was associated with the sun or the Thirumal in the mythology of ancient Tamil religion. Yellow is the colour of the solar plexus chakra, which in traditional Tamil Siddha medicine is the energy centre relating to the metabolic and digestive systems. Orange is the colour of the sacral chakra, and tied to the reproductive system.

The plant is used in poosai to represent a form of Kottravai who is said to reside on this plant as herself. The plant is used as a component of navapatrika along with plantain (Banana), Kachvi or Kacci or kochu or Taro, jayanti/ Barley, wood apple (Bilva), pomegranate (darimba), Asoka, manaka or Manakochu and rice paddy or Dhanya.

It is used in poosai to make a form of Ganesha. Yaanaimugathaan, the remover of obstacles, is invoked at the beginning of almost any ceremony and a form of Yaanaimugathaan for this purpose is made by mixing turmeric with water and forming it into a cone-like shape.

Gaye holud (literally "yellow on the body") is a ceremony observed mostly in the region of Bengal (comprising Bangladesh and Indian West Bengal). The gaye holud takes place one or two days before the religious and legal Bengali wedding ceremonies. The turmeric paste is applied by friends to the bodies of the couple. This is said to soften the skin, but also colours them with the distinctive yellow hue that gives its name to this ceremony. It may be a joint event for the bride and groom's families, or it may consist of separate events for the bride's family and the groom's family.

During the Tamil festival Pongal, a whole turmeric plant with fresh rhizomes is offered as a thanksgiving offering to Suryan, the Sun god. Also, the fresh plant sometimes is tied around the sacred Pongal pot in which an offering of pongal is prepared.

In Tamil nadu, as a part of the Tamil marriage ritual, dried turmeric tuber tied with string is used temporarily or permanently as opposed to the Mangalasutra of Hindus in India . The Tamil Marriage act recognizes this custom. Thali necklace is the equivalent of marriage rings in western cultures. In western and coastal India, during weddings of the Marathi and Konkani people, Kannada Brahmins turmeric tubers are tied with strings by the couple to their wrists during a ceremony called *Kankanabandhana.*

Friedrich Ratzel in *The History of Mankind* reported in 1896 that in Micronesia the preparation of turmeric powder for embellishment of body, clothing and utensils had a highly ceremonial character. He quotes an example of the roots being ground by four to six women in special public buildings and then allowed to stand in water. The following morning, three young coconuts and three old soma nuts are offered by a priestess with prayer, after which the dye which has settled down in the water is collected, baked into cakes in coconut molds, wrapped in banana leaves, and hung up in the huts till required for use.

Ginger

Ginger or ginger root is the rhizome of the plant *Zingiber officinale,* consumed as a delicacy, medicine, or spice. It lends its name to its genus and family (Zingiberaceae). Other notable members of this plant family are turmeric, cardamom, and galangal. The distantly related dicots in the Asarum genus have the common name wild ginger because of their similar taste. Ginger is indigenous to southern China, from whence it is spread to the Spice Islands and other parts of Asia, and subsequently to West Africa and to the Caribbean.

Horticulture

Ginger produces clusters of white and pink flower buds that bloom into yellow flowers. Because of its aesthetic appeal and the adaptation of the plant to warm climates, ginger is often used as landscaping around subtropical homes. It is a perennial reed-like plant with annual leafy stems, about a metre (3 to 4 feet) tall. Traditionally, the rhizome is gathered when the stalk withers; it is immediately scalded, or washed and scraped, to kill it and prevent sprouting. The fragrant perisperm of Zingiberaceae is used as sweetmeats by Bantu, also as a condiment and sialogogue.

Uses

Ginger produces a hot, fragrant kitchen spice. Young ginger rhizomes are juicy and fleshy with a very mild taste. They are often pickled in vinegar or sherry as a snack or just cooked as an ingredient in many dishes. They can also be steeped in boiling water to make ginger tea, to which honey is often added; sliced orange or lemon fruit may also be added. Ginger can also be made into candy, or ginger wine which has been made commercially since 1740.

Mature ginger rhizomes are fibrous and nearly dry. The juice from old ginger roots is extremely potent and is often used as a spice in Indian recipes, and is a quintessential ingredient of Chinese, Korean, Japanese and many South Asian cuisines for flavoring dishes such as seafood or goat meat and vegetarian cuisine.

Ginger acts as a useful food preservative. Fresh ginger can be substituted for ground ginger at a ratio of 6 to 1, although the flavors of fresh and dried ginger are somewhat different. Powdered dry ginger root is typically used as a flavoring for recipes such as gingerbread, cookies, crackers and cakes, ginger ale, and ginger beer.

Candied ginger, or crystallized ginger, is the root cooked in sugar until soft, and is a type of confectionery.

Fresh ginger may be peeled before eating. For longer-term storage, the ginger can be placed in a plastic bag and refrigerated or frozen.

Regional Use

In Western cuisine, ginger is traditionally used mainly in sweet foods such as ginger ale, gingerbread, ginger snaps, parkin, ginger biscuits and speculaas. A ginger-flavored liqueur called Canton is produced in Jarnac, France. Ginger wine is a ginger-flavored wine produced in the United Kingdom, traditionally sold in a green glass bottle. Ginger is also used as a spice added to hot coffee and tea.

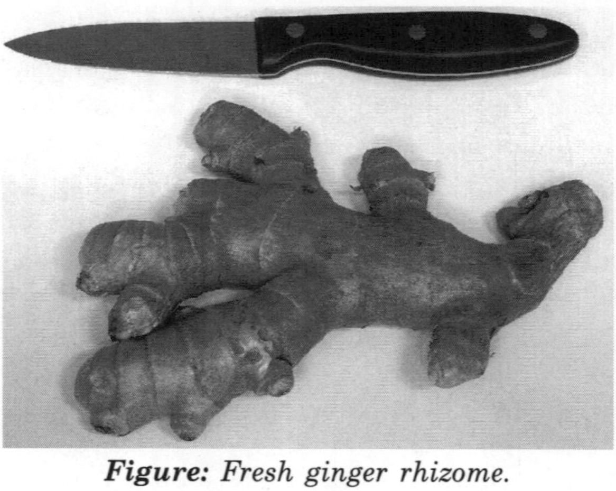

Figure: *Fresh ginger rhizome.*

In India and Pakistan, ginger is called *adrak* in Hindi, Punjabi and Urdu, *aad* in Maithili. Fresh ginger is one of the main spices used for making pulse and lentil curries and other vegetable preparations. Fresh, as well as dried, ginger is used to spice tea and coffee, especially in winter. Ginger powder is also used in certain food preparations, particularly for pregnant or nursing women, the most popular one being *katlu* which is a mixture of gum resin, ghee, nuts, and sugar. Ginger is also consumed in candied and pickled form. In Bangladesh, ginger is finely chopped or ground into a paste to use as a base for chicken and meat dishes alongside onion and garlic.

In Burma, ginger is called *gyin*. It is widely used in cooking and as a main ingredient in traditional medicines. It is also consumed as a salad dish called *gyin-thot*, which consists of shredded ginger preserved in oil, and a variety of nuts and seeds. In Indonesia, a beverage called *wedang jahe* is made from ginger and palm sugar. Indonesians also use ground ginger root, called *jahe*, as a common ingredient in local recipes. In Malaysia, ginger is called *halia* and used in many kinds of dishes, especially a soup. In the Philippines, it is a common ingredient in local dishes and it is brewed into a tea called *salabat*. In Vietnam, the fresh leaves, finely chopped, can also be added to shrimp-and-yam soup (*canh khoai mú*) as a top garnish and spice to add a much subtler flavour of ginger than the chopped root.

In China, sliced or whole ginger root is often paired with savory dishes such as fish, and chopped ginger root is commonly paired with meat, when it is cooked. However, candied ginger is sometimes a component of Chinese candy boxes, and an herbal tea can also be prepared from ginger. In Japan, ginger is pickled to make beni shoga and gari or grated and used raw on tofu or noodles. It is also made into a candy called *shoga no sato zuke*. In the traditional Korean kimchi, ginger is either finely minced or just juiced in order to avoid the fibrous texture and added to the ingredients of the spicy paste just before the fermenting process.

In the Caribbean, ginger is a popular spice for cooking, and making drinks such as *sorrel*, a seasonal drink made during the Christmas season. Jamaicans make ginger beer both as a carbonated beverage and also fresh in their homes. Ginger tea is often made from fresh ginger, as well as the famous regional speciality Jamaican ginger cake.

On the island of Corfu, Greece, a traditional drink called *tsitsibira*, a type of ginger beer, is made. The people of Corfu and the rest of the Ionian islands adopted the drink from the British, during the period of the United States of the Ionian Islands.

In Arabic, ginger is called *zanjabil*, and in some parts of the Middle East. From its main ingredient ginger tea derives a flavour that is spicy and stimulating. Ginger, known as Adarak in Hindi, is used frequently in tea made in all parts of India as well.

Medicinal Use and Research

According to the American Cancer Society, ginger has been promoted as a cancer treatment "to keep tumors from developing", but "available scientific evidence does not support this". They add: "Recent preliminary results in animals show some effect in slowing or preventing tumour growth. While these results are not well understood, they deserve further study. Still, it is too early in the research process to say whether ginger will have the same effect in humans."

In limited studies, ginger was found to be more effective than placebo for treating nausea caused by seasickness, morning sickness and chemotherapy, although ginger was not found superior to placebo for pre-emptively treating post-operative nausea. Some studies advise against taking ginger during pregnancy, suggesting that ginger is mutagenic, though some other studies have reported antimutagenic effects. Other preliminary studies showed that ginger may affect arthritis pain or have blood thinning and cholesterol lowering properties, but these effects remain unconfirmed.

A 2013 *in vivo* evaluation demonstrated ginger extract showed a hepatoprotective effect in rats. A 2013 review found that ginger is a free radical scavenger, antioxidant; thus inhibits lipid peroxidation and that these attributes could be contributing to its known gastroprotective effects. A 2012 review found ginger extract and ginger juice possess anti-emetic effects against chemotherapy-induced nausea and vomiting in experimental animals. A 2012 review found the radioprotective properties of ginger extract might be effective to protect against gamma radiation-induced side effects from cancer treatment in mice. A 2011 review found ginger displays chemopreventive and antineoplastic effects. The same review found that ginger appears to be promising for cancer prevention, though further research is necessary to evaluate the efficacy and safety of ginger. Advanced glycation end-products are possibly associated in the development of diabetic cataract for which ginger was effective in preliminary studies, apparently by acting through antiglycating mechanisms. Zingerone may have activity against enterotoxigenic *Escherichia coli* in enterotoxin-induced diarrhea in mice.

Chemistry

The characteristic odor and flavour of ginger is caused by a mixture of zingerone, shogaols and gingerols, volatile oils that compose one to three percent of the weight of fresh ginger. In laboratory animals, the gingerols increase the motility of the gastrointestinal tract and have analgesic, sedative, antipyretic and antibacterial properties. A study at the University of Michigan demonstrated that gingerols can inhibit growth of ovarian cancer cells *in vitro*. [6]-gingerol (1-[4'-hydroxy-3'-methoxyphenyl]-5-hydroxy-3-decanone) is the major pungent principle of ginger. Ginger contains up to three percent of a fragrant essential oil whose main constituents are sesquiterpenoids, with (-)-zingiberene as the main component. Smaller amounts of other sesquiterpenoids (â-sesquiphellandrene, bisabolene and farnesene) and a small monoterpenoid fraction (â-phelladrene, cineol, and citral) have also been identified.

The pungent taste of ginger is due to nonvolatile phenylpropanoid-derived compounds, particularly gingerols and shogaols, which form from gingerols when ginger is dried or cooked. Zingerone is also produced from gingerols during this process; this compound is less pungent and has a spicy-sweet aroma. Ginger is also a minor chemical irritant, and because of this was used as a horse suppository by pre-World War I mounted regiments for feaguing.

Ginger has a sialagogue action, stimulating the production of saliva, which makes swallowing easier.

Folk Medicine

The traditional medical form of ginger historically was called *Jamaica ginger*; it was classified as a stimulant and carminative and used frequently for dyspepsia, gastroparesis, slow motility symptoms, constipation, and colic. It was also frequently employed to disguise the taste of medicines. Some studies indicate ginger may provide short-term relief of pregnancy-related nausea and vomiting. Studies are inconclusive about effects for other forms of nausea or in treating pain from rheumatoid arthritis, osteoarthritis, or joint and muscle injury. Side effects, mostly associated with powdered ginger, are gas, bloating, heartburn, and nausea.

Tea brewed from ginger is a common folk remedy for colds. Ginger ale and ginger beer are also drunk as *stomach settlers* in countries where the beverages are made.

- In Burma, ginger and a local sweetener made from palm tree juice (*htan nyat*) are boiled together and taken to prevent the flu.
- In China, ginger is included in several traditional preparations. A drink made with sliced ginger cooked in water with brown sugar or a cola is used as a folk medicine for the common cold. "Ginger eggs" (scrambled eggs with finely diced ginger root) is a common home remedy for coughing. The Chinese also make a kind of dried ginger candy that is fermented in plum juice and sugared, which is also commonly consumed to suppress coughing.

Ginger has also been historically used to treat inflammation, which several scientific studies support, though one arthritis trial showed ginger to be no better than a placebo or ibuprofen for treatment of osteoarthritis.

- In Congo, ginger is crushed and mixed with mango tree sap to make tangawisi juice, which is considered a panacea.
- In India, ginger is applied as a paste to the temples to relieve headache, and consumed when suffering from the common cold. Ginger with lemon and black salt is also used for nausea.
- In Indonesia, ginger (*jahe* in Indonesian) is used as a herbal preparation to reduce fatigue, reducing "winds" in the blood, prevent and cure rheumatism and control poor dietary habits.
- In the Philippines, ginger is known as *luya* and is used as a throat lozenge in traditional medicine to relieve sore throat. It is also brewed into a tea known as *salabat*.
- In the United States, ginger is used to prevent motion and morning sickness. It is recognized as safe by the Food and Drug Administration and is sold as an unregulated dietary supplement. Ginger water is also used to avoid heat cramps.
- In Peru, ginger is sliced in hot water as an infusion for stomach aches as *infusión de Kión*.
- In Japan it is purported to aid blood circulation. Scientific studies investigating these effects have been inconclusive.

Consumed in reasonable quantities, ginger has few negative side effects, and is on the FDA's "generally recognized as safe" list, though it does interact with some medications, including warfarin. Ginger is contraindicated in people suffering from gallstones, because it promotes the production of bile.

Allergic reactions to ginger generally result in a rash, and although generally recognized as safe, ginger can cause heartburn, bloating, gas, belching and nausea, particularly if taken in powdered form. Unchewed fresh ginger may result in intestinal blockage, and individuals who have had ulcers, inflammatory bowel disease or blocked intestines may react badly to large quantities of fresh ginger. Ginger can also adversely affect individuals with gallstones. There are also suggestions that ginger may affect blood pressure, clotting, and heart rhythms.

Products in Taiwan made from Hebo Natural Products Limited of China contained ginger contaminated with DIBP, some 80,000 nutritional supplement capsules made with imported ginger powder were seized by the Public Health Department of Taiwan in June 2011.

Similar Ingredients

Myoga (*Zingiber mioga* Roscoe) appears in Japanese cuisine; the flower buds are the part eaten. Another plant in the *Zingiberaceae* family, galangal, is used for similar purposes as ginger in Thai cuisine. Galangal is also called Thai ginger, fingerroot (*Boesenbergia rotunda*), Chinese ginger, or the Thai *krachai*. A dicotyledonous native species of eastern North America, *Asarum canadense*, is also known as "wild ginger", and its root has similar aromatic properties, but it is not related to true ginger. The plant also contains aristolochic acid, a carcinogenic compound.

Onion

The onion (*Allium cepa*) (Latin 'cepa' = onion), also known as the bulb onion or common onion, is used as a vegetable and is the most widely cultivated species of the genus *Allium*. This genus also contains several other species variously referred to as onions and cultivated for food, such as the Japanese bunching onion (*A. fistulosum*), the Egyptian onion (*A.* ×*proliferum*), and the Canada onion (*A. canadense*). The name "wild onion" is applied to a number of *Allium* species but *A. cepa* is exclusively known from cultivation and its ancestral wild original form is not known, although escapes from cultivation have become established in some regions. The onion is most frequently a biennial or a perennial plant, but is usually treated as an annual and harvested in its first growing season.

The onion plant has a fan of hollow, bluish-green leaves and the bulb at the base of the plant begins to swell when a certain day-length is reached. In the autumn the foliage dies down and the outer layers of the bulb become dry and brittle. The crop is harvested and dried and the onions are ready for use or storage. The crop is prone to attack by a number of pests and diseases, particularly the onion fly, the onion eelworm and various fungi that cause rotting. Some varieties of *A. cepa* such as shallots and potato onions produce multiple bulbs. Onions are cultivated and used around the world. As a foodstuff they are usually served cooked, as a vegetable or part of a prepared savoury dish, but can also be eaten raw or used to make pickles or chutneys. They are pungent when chopped and contain certain chemical substances which irritate the eyes. Onions contain phenolics and flavonoids that have potential anti-inflammatory, anti-cholesterol, anticancer and antioxidant properties.

Description

The onion plant (*Allium cepa*) is unknown in the wild but has been grown and selectively bred in cultivation for at least 7,000 years. It is a biennial plant but is usually grown as an annual. Modern varieties typically grow to a height of 15 to 45 cm (6 to 18 in). The leaves are blueish-green and grow alternately in a flattened, fan-shaped swathe. They are fleshy, hollow and cylindrical, with one flattened side. They are at their broadest about a quarter of the way up beyond which they taper towards a blunt tip. The base of each leaf is a flattened, usually white sheath that grows out of a basal disc. From the underside of the disc, a bundle of fibrous roots extends for a short way into the soil. As the onion matures, food reserves begin to accumulate in the leaf bases and the bulb of the onion swells.

In the autumn the leaves die back and the outer scales of the bulb become dry and brittle, and this is the time at which the crop is normally harvested. If left in the soil over winter, the growing point in the middle of the bulb begins to develop in the spring. New leaves appear and a long, stout, hollow stem expands, topped by a bract protecting a developing inflorescence. The inflorescence takes the form of a globular umbel of white flowers with parts in sixes. The seeds are glossy black and triangular in cross section.

Uses

Historical Use: Bulbs from the onion family are thought to have been used as a food source for millennia. In Bronze Age settlements, traces of onion remains were found alongside date stones and fig remains that date back to 5000 BC. However, it is not clear if these were cultivated onions. Archaeological and literary evidence such as the Book of Numbers 11:5 suggests that onions were probably being cultivated around two thousand years later in ancient Egypt, at the same time that leeks and garlic were cultivated. Workers who built the Egyptian pyramids may have been fed radishes and onions.

Figure: *A red onion.*

The onion is easily propagated, transported and stored. The ancient Egyptians worshipped it, believing its spherical shape and concentric rings symbolized eternal life. Onions were even used in Egyptian burials, as evidenced by onion traces being found in the eye sockets of Ramesses IV.

In ancient Greece, athletes ate large quantities of onion because it was believed to lighten the balance of the blood. Roman gladiators were rubbed down with onions to firm up their muscles. In the Middle Ages, onions were such an important food that people would pay their rent with onions, and even give them as gifts. Doctors were known to prescribe onions to facilitate bowel movements and erections, and to relieve headaches, coughs, snakebite and hair loss.

Onions were taken by the first settlers to North America, where the Native Americans were already using wild onions in a number of ways, eating them raw or cooked in a variety of foods. They also used them to make into syrups, to form poultices and in the preparation of dyes. According to diaries kept by the colonists, bulb onions were one of the first things planted by the Pilgrim Fathers when they cleared the land for cropping in 1648.

Onions were also prescribed by doctors in the early 16th century to help with infertility in women. They were similarly used to raise fertility levels in dogs, cats and cattle, but this was an error as recent research has shown that onions are toxic to dogs, cats, guinea pigs and many other animals.

Culinary Uses

Onions are often chopped and used as an ingredient in various hearty warm dishes, and may also be used as a main ingredient in their own right, for example in French onion soup or onion chutney. They are very versatile and can be baked, boiled, braised, grilled, fried, roasted, sautéed or eaten raw in salads. Their layered nature makes them easy to hollow out once cooked, facilitating stuffing them. Onions are a staple in Indian cuisine, often used as a thickening agent for curries and gravies. Onions pickled in vinegar are eaten as a snack. These are often served as a side serving in pubs and fish and chip shops throughout the United Kingdom and The Commonwealth, often served with cheese and/or ale in the United Kingdom. In North America, sliced onions are battered and deep fried and served as onion rings.

Onion Types and Products

Common onions are normally available in three colour varieties. *Yellow* or *brown* onions (called *red* in some European countries), are full-flavoured and are the onions of choice for everyday use. Yellow onions turn a rich, dark brown when caramelized and give French onion soup its sweet flavour. The *red* onion (called *purple* in some European countries) is a good choice for fresh use when its colour livens up the dish. It is also used in grilling and char-broiling. *White* onions are the traditional onions that are used in classic Mexican cuisine. They have a golden colour when cooked and a particularly sweet flavour when sautéed.

While the large mature onion bulb is the onion most often eaten, onions can be eaten at immature stages. Young plants may be harvested before bulbing occurs and used whole as scallions. When an onion is harvested after bulbing has begun but the onion is not yet mature, the plants are sometimes referred to as *summer* onions.

Additionally, onions may be bred and grown to mature at smaller sizes. Depending on the mature size and the purpose for which the onion is used, these may be referred to as *pearl*, *boiler*, or *pickler* onions, but differ from true pearl onions which are a different species. Pearl and boiler onions may be cooked as a vegetable rather than as an ingredient and pickler onions are often preserved in vinegar as a long-lasting relish.

Onions are available in fresh, frozen, canned, caramelized, pickled and chopped forms. The dehydrated product is available as kibbled, sliced, rings, minced, chopped, granulated and powder forms. Onion powder is a spice widely used when the fresh ingredient is not available. It is made from finely ground, dehydrated onions, mainly the pungent varieties of bulb onions, and has a strong odour. Being dehydrated, it has a long shelf life and comes in several varieties: yellow, red and white.

Non-culinary Uses

Onions have particularly large cells that are readily observed by the human eye at low magnification; consequently, onion tissue is frequently used in science education for demonstrating microscope usage and for learning about the structure of cells.

The pungent juice of onions has been used as a moth repellent and can be rubbed on the skin to prevent insect bites. When applied to the scalp it is said to promote growth of hair and on the face to reduce freckling. It has been used to polish glass and copperware and to prevent rust on iron. If boiling water is poured onto chopped onions and left to cool, the resulting liquor can be sprayed onto plants to increase their resistance to pests, and the onion plants when growing are reputed to keep away moles and insects. Onion skins have been used to produce a yellow-brown dye.

Nutrition and Health

Most onion cultivars are about 89% water, 4% sugar, 1% protein, 2% fibre and 0.1% fat. They contain vitamin C, vitamin B_6, folic acid and numerous other nutrients in small amounts. They are low in fats and in sodium, and with an energy value of 166kJ (40 kcal) per 100 g (3.5 oz) serving, they can contribute their flavour to savoury dishes without raising caloric content appreciably.

Onions contain chemical compounds such as phenolics and flavonoids that basic research shows to have potential anti-inflammatory, anti-cholesterol, anticancer and antioxidant properties. These include quercetin and its glycosides quercetin 3,4'-diglucoside and quercetin-4'-glucoside. There are considerable differences between different varieties in potential antioxidant content. Shallots have the highest level, six times the amount found in Vidalia onions, the variety with the smallest amount.

Some people suffer from allergic reactions after handling onions. Symptoms can include contact dermatitis, intense itching, rhinoconjunctivitis, blurred vision, bronchial asthma, sweating and anaphylaxis. There may be no allergic reaction in these individuals to the consumption of onions, perhaps because of the denaturing of the proteins involved during the cooking process.

While onions and other members of the genus *Allium* are commonly consumed by humans, they can be deadly for dogs, cats, guinea pigs, monkeys and other animals. The toxicity is caused by the sulfoxides present in raw and cooked onions which many animals are unable to digest. Ingestion results in anaemia caused by the distortion and rupture of red blood cells. Sick pets are sometimes fed with tinned baby foods and any that contain onion should be avoided. Nor is it good for pets to be fed onion-containing leftovers such as pizza, canned spaghetti, Chinese dishes and onion rings. The typical toxic doses are 5 g (0.2 oz) per kg (2.2 lb) bodyweight for cats and 15 to 30 g (0.5 to 1.1 oz) per kg for dogs.

In India, some sects do not eat onions as they believe them to be an aphrodisiac. Various schools of Buddhism also advise against the consumption of onions and garlic because they believe that they increase desire when eaten cooked and anger when eaten raw.

Eye Irritation

Chopping an onion causes damage to cells which allows enzymes called alliinases to break down amino acid sulfoxides and generate sulfenic acids. A specific sulfenic acid, 1-propenesulfenic acid, is rapidly acted on by a second enzyme, the lachrymatory factor synthase (LFS), giving syn-propanethial-S-oxide, a volatile gas known as the onion lachrymatory factor or LF. This gas diffuses through the air and soon reaches the eye, where it activates sensory neurons, creating a stinging sensation. Tear glands produce tears in order to dilute and flush out the irritant. Eye irritation can be avoided by cutting onions under running water or submerged in a basin of water. Leaving the root end intact also reduces irritation as the onion base has a higher concentration of sulphur compounds than the rest of the bulb. Refrigerating the onions before use reduces the enzyme reaction rate and using a fan can blow the gas away from the eyes. The more often one chops onions, the less one experiences eye irritation.

The amount of sulfenic acids and LF released and the irritation effect differs among *Allium* species. In 2008, the New Zealand Crop and Food institute created a strain of "no tears" onions by using gene-silencing biotechnology to prevent synthesis by the onions of the LFS enzyme.

Cultivation

Onions are best cultivated in fertile soils that are well-drained. Sandy loams are good as they are low in sulphur, while clayey soils usually have a high sulphur content and produce pungent bulbs. Onions require a high level of nutrients in the soil. Phosphorus is often present in sufficient quantities but may be applied before planting because of its low level of availability in cold soils. Nitrogen and potash can be applied at intervals during the growing season, the last application of nitrogen being at least four weeks before harvesting. Bulbing onions are day-length sensitive; their bulbs begin growing only after the number of daylight hours has surpassed some minimal quantity. Most traditional European onions are what is referred to as "long-day" onions, producing bulbs only after 14+ hours of daylight occurs. Southern European and North African varieties are often known as "intermediate day" types, requiring only 12–13 hours of daylight to stimulate bulb formation. Finally, "short-day" onions, which have been developed in more recent times, are planted in mild-winter areas in the fall and form bulbs in the early spring, and require only 11–12 hours of daylight to stimulate bulb formation. Onions are a cool-weather crop and can be grown in USDA zones 3 to 9. Hot temperatures or other stressful conditions cause them to "bolt", meaning that a flower stem begins to grow.

Onions may be grown from seed or from sets. Onion seeds are short lived and fresh seed germinates better. The seeds are sewn thinly in shallow drills, thinning the plants in stages. In suitable climates, certain cultivars can be sown in late summer and autumn to overwinter in the ground and produce early crops the following year. Onion sets are produced by sowing seed thickly in early summer in poor soil and the small bulbs produced are harvested in the autumn. These bulbs are planted the following spring and grow into mature bulbs later in the year. Certain cultivars are used for this purpose and these may not have such good storage characteristics as those grown directly from seed.

Routine care during the growing season involves keeping the rows free of competing weeds, especially when the plants are young. The plants are shallow rooted and do not need a great deal of water when established. Bulbing usually takes place after twelve to eighteen weeks. The bulbs can be gathered when needed to eat fresh but if required to be kept in store, should be harvested after the leaves have died back naturally. In dry weather they can be left on the surface of the soil for a few days to dry out properly, then they can be placed in nets, roped into strings or laid in layers in shallow boxes. They should be stored in a well-ventilated, cool place such as a shed.

Pests and Diseases

Onions suffer from a number of plant disorders. The most serious for the home gardener are likely to be the onion fly, stem and bulb eelworm, white rot and neck rot. Diseases affecting the foliage include rust and smut, downy mildew and white tip disease. The bulbs may be affected by splitting, white rot and neck rot. Shanking is a condition when the central leaves turn yellow and the inner part of the bulb collapses into an unpleasant-smelling slime. Most of these disorders are best treated by removing and burning affected plants. The larvae of the onion leaf miner or leek moth (*Acrolepiopsis assectella*) sometimes attacks the foliage and may burrow down into the bulb.

The onion fly (*Delia antiqua*) lays eggs on the leaves and stems and on the ground close to onion, shallot, leek and garlic plants. The fly is attracted to the crop by the smell of damaged tissue and is liable to occur after thinning. Plants grown from sets are less prone to attack. The larvae tunnel into the bulbs and the foliage wilts and turns yellow. The bulbs are disfigured and rot, especially in wet weather. Control measures may include crop rotation, the use of seed dressings, early sowing or planting and the removal of infested plants.

The onion eelworm (*Ditylenchus dipsaci*), a tiny parasitic soil-living nematode, causes swollen distorted foliage. Young plants are killed and older ones produce soft bulbs. There is no cure and affected plants should be uprooted and burnt. The site should not be used for growing onions again for several years and should also be avoided for growing carrots, parsnips and beans which are also susceptible to the eelworm. White rot of onions, leeks and garlic is caused by the soil-borne fungus *Sclerotium cepivorum*. As the roots rot, the foliage turns yellow and wilts. The base of the bulbs is attacked and becomes covered by a fluffy white mass of mycelia which later produces small, globular black structures called sclerotia. These resting structures remain in the soil to reinfect a future crop. There is no cure for this fungal disease so affected plants should be removed and destroyed and the ground used for unrelated crops in subsequent years.

Neck rot is a fungal disease affecting onions in storage. It is caused by *Botrytis allii* which attacks the neck and upper parts of the bulb causing a grey mould to develop. The symptoms often first occur where the bulb has been damaged and spread downwards in the affected scales. Large quantities of spores are produced and crust-like sclerotia may also develop. In time a dry rot sets in and the bulb becomes a dry, mummified structure. This disease may be present throughout the growing period but only manifest itself when the bulb is in store. Anti-fungal seed dressings are available and the disease can be minimised by preventing physical damage to the bulbs at harvesting, careful drying and curing of the mature onions and correct storage in a cool, dry place with plenty of circulating air.

Storage in the Home

Cooking onions and sweet onions are better stored at room temperature, optimally in a single layer, in mesh bag in a dry, cool, dark, well-ventilated location. In this environment, cooking onions have a shelf life of three to four weeks and sweet onions one to two weeks. Cooking onions will absorb odours from apples and pears. Also, they draw moisture from vegetables with which they are stored which may cause them to decay.

Sweet onions have a greater water and sugar content than cooking onions. This makes them sweeter and milder tasting but reduces their shelf life. Sweet onions can be stored refrigerated; they have a shelf life of approximately one month. Irrespective of type, any cut pieces of onion are best tightly wrapped, stored away from other produce, and used within two to three days.

Varieties

Common Onion Group (var. Cepa): Most of the diversity within *A. cepa* occurs within this group, the most economically important *Allium* crop. Plants within this group form large single bulbs, and are grown from seed or seed-grown sets. The majority of cultivars grown for dry bulbs, salad onions, and pickling onions belong to this group. The range of diversity found among these cultivars includes variation in photoperiod (length of day that triggers bulbing), storage life, flavour, and skin colour. Common onions range from the pungent varieties used for dried soups and onion powder to the mild and hearty sweet onions, such as the Vidalia from Georgia, USA, or Walla Walla from Washington that can be sliced and eaten on a sandwich instead of meat.

Aggregatum Group (var. Aggregatum): This group contains shallots and potato onions, also referred to as multiplier onions. The bulbs are smaller than those of common onions, and a single plant forms an aggregate cluster of several bulbs. They are propagated almost exclusively from daughter bulbs, although reproduction from seed is possible. Shallots are the most important subgroup within this group and comprise the only cultivars cultivated commercially. They form aggregate clusters of small, narrowly ovoid to pear-shaped bulbs. Potato onions differ from shallots in forming larger bulbs with fewer bulbs per cluster, and having a flattened (onion-like) shape. However, intermediate forms exist.

I'itoi onion (*Allium cepa*) is a prolific multiplier onion cultivated in the Baboquivari Peak Wilderness, Arizona area. This small bulb type has a shallot-like flavour and is easy to grow and ideal for hot, dry climates. Bulbs are separated, and planted in the fall 1 inch below surface and 12 inches apart. Bulbs will multiply into clumps and can be harvested throughout the cooler months. Tops will die back in the heat of summer and may return with heavy rains; bulbs can remain in the ground or be harvested and stored in a cool dry place for planting in the fall. The plants rarely flower; propagation is by division.

Hybrids with A. Cepa Parentage

A number of hybrids are cultivated that have *A. cepa* parentage, such as the diploid tree onion or Egyptian onion (*A.* ×*proliferum*), Wakegi onion (*A.* ×*wakegi*), and the triploid onion (*A.* ×*cornutum*).

The tree onion or Egyptian onion produces bulblets in the umbel instead of flowers, and is now known to be a hybrid of *A. cepa* × *A. fistulosum*. It has previously been treated as a variety of *A. cepa*, for example *A. cepa* var. *proliferum*, *A. cepa* var. *bulbiferum*, and *A. cepa* var. *viviparum*.

The Wakegi onion is also known to be a hybrid between *A. cepa* and *A. fistulosum*, with the *A. cepa* parent believed to be from the Aggregatum Group of cultivars. It has been grown for centuries in Japan and China for use as a salad onion.

Under the rules of botanical nomenclature, both the Egyptian onion and Wakegi onion should be combined into one hybrid species, having the same parent species. Where this is followed, the Egyptian onion is named *A. ×proliferum* Eurasian Group and the Wakegi onion is named *A. ×proliferum* East Asian Group.

The triploid onion is a hybrid species with three sets of chromosomes, two sets from *A. cepa* and the third set from an unknown parent. Various clones of the triploid onion are grown locally in different regions, such as 'Ljutika' in Croatia, and 'Pran', 'Poonch' and 'Srinagar' in the India-Kashmir region. 'Pran' is grown extensively in the Northern Indian provinces of Jammu and Kashmir. There are very small genetic differences between 'Pran' and the Croatian clone 'Ljutika', implying a monophyletic origin for this species.

Some authors have used the name *A. cepa* var. *viviparum* (Metzg.) Alef. for the triploid onion, but this name has also been applied to the Egyptian onion. The only name unambiguously connected with the triploid onion is *A. ×cornutum*.

Scallions or salad onions may be grown from the Welsh onion (*A. fistulosum*) as well as from *A. cepa*. Young plants of *A. fistulosum* and *A. cepa* look very similar, but may be distinguished by their leaves, which are circular in cross-section in *A. fistulosum* rather than flattened on one side.

Production and Trade

It is estimated that around the World, over 9,000,000 acres (3,642,000 ha) of onions are grown annually. About 170 countries cultivate onions for domestic use and about eight percent of the global production is traded internationally.

Asafoetida

Asafoetida or asafetida (*Ferula assa-foetida*) /æsYÈf[thdY/, is the dried latex (gum oleoresin) exuded from the rhizome or tap root of several species of *Ferula*, a perennial herb that grows 1 to 1.5 m tall. The species is native to the deserts of Iran, mountains of Afghanistan, and is mainly cultivated in nearby India. As its name suggests, asafoetida has a fetid smell but in cooked dishes it delivers a smooth flavour reminiscent of leeks. It is also known as asant, food of the gods, giant fennel, jowani badian, stinking gum, Devil's dung, hing, kayam and ting.

Uses

Cooking: This spice is used as a digestive aid, in food as a condiment, and in pickles. It typically works as a flavour enhancer and, used along with turmeric, is a standard component of Indian cuisine, particularly in lentil curries, such as dal, or various 'Pappu' like dosakaya (cucumber), or tomato pappu, mamidikaya (raw mango) in Telugu, called Mavinikaya in Kannada. as well as in numerous vegetable dishes. It is especially widely used in South Indian and Maharashtrian cuisine, which is mainly vegetarian, and is often used to harmonize sweet, sour, salty and spicy components in food. It is used to hallmark the taste of Andhra Pradesh, Karnataka, Kerala and Tamil Nadu sambar, a saucy dish made with cereals and lentils. The spice is added to the food at the time of Chaunk / Popu/ tadka (tempering). Sometimes dried and ground asafoetida (in very mild quantity) can be mixed with salt and eaten with raw salad. In its pure form, its odour is so strong the aroma will contaminate other spices stored nearby if it is not stored in an airtight container: many commercial preparations of asafoetida utilize the resin ground up and mixed with a larger volume of wheat flour: the mixture is sold in sealed plastic containers with a small hole at the bottom, allowing the diluted spice to be dusted lightly over the food being cooked. However, its odour and flavour become much milder and more pleasant upon heating in oil or ghee, acquiring a taste and aroma reminiscent of sautéed onion and garlic. It is used especially by the merchant caste of the Hindus and by adherents of Jainism and Vaishnavism, particularly in Rajasthan, Gujarat and Maharashtra, who do not eat onions or garlic. It is used in many vegetarian and lentil dishes to add both flavour and aroma as well as to reduce flatulence. It is however one of the pungent vegetables generally avoided by Buddhist vegetarians.

Medical Applications

- Antiflatulent. Asafoetida reduces the growth of indigenous microflora in the gut, reducing flatulence. In the Jammu region of India, asafoetida is used as a medicine for flatulence and constipation by 60% of locals.
- A digestion aid. In Thailand and India, it is used to aid digestion and is smeared on the abdomen in an alcohol or water tincture known as *mahahing*. Assafoetida in this tincture form was evidently used in western medicine as a topical treatment for abdominal injuries during the 18th and 19th centuries, although when it came into use in the West and how long it remained in use is uncertain. One notable case in which it was used is that of Canadian *Coureur des bois* Alexis St. Martin, who in 1822 suffered a severe abdominal injury from an accidental shooting that perforated his right lung and stomach and shattered several

ribs. St Martin was treated by American army surgeon William Beaumont, who subsequently used St Martin as the subject of a pioneering series of experiments in gastric physiology, thanks to the fact that when St Martin's wounds fully healed, it left an open fistula into the stomach that enabled Beaumont to insert various types of food directly into St Martin's stomach and record the results. In his account of his treatment of and later experiments on St Martin, Beaumont recorded that he treated the suppurating chest wound with a combination of wine mixed with diluted muriatic acid and 30-40 drops of tincture of asafoetida applied three times a day, and that this appeared to have the desired effect, helping the wound to heal.

- Fighting influenza: Asafoetida was used in 1918 to fight the Spanish influenza pandemic. In 2009, researchers reported that the roots of Asafoetida produce natural antiviral drug compounds that demonstrated potency against the H1N1 virus in vitro and concluded that "sesquiterpene coumarins from *F. assa-foetida* may serve as promising lead compounds for new drug development against influenza A (H1N1) viral infection".
- Remedy for asthma and bronchitis. It is also said to be helpful in cases of asthma and bronchitis. A folk tradition remedy for children's colds: it is mixed into a pungent-smelling paste and hung in a bag around the afflicted child's neck.
- An antimicrobial: Asafoetida has a broad range of uses in traditional medicine as an antimicrobial, with well documented uses for treating chronic bronchitis and whooping cough, as well as reducing flatulence.
- A contraceptive/abortifacient: Asafoetida has also been reported to have contraceptive/abortifacient activity,. It is related to (and considered an inferior substitute for) the ancient *Ferula* species Silphium.
- Antiepileptic: Asafoetida oleo-gum-resin has been reported to be antiepileptic in classical Unani, as well as ethnobotanical literature.
- Balancing the *vata* and *kapha*. In India according to the Ayurveda, asafoetida is considered to be one of the best spices for balancing the *vata dosha. It mitigates vata and kapha, relieves flatulence and colic pain. It is pungent in taste and at the end of digestion. It aggravates pitta, enhances appetite, taste and digestion. It is easy to digest.* (ref: ashtanga hridaya Su chapter 6).

- Antidote for opium. Asafoetida has only been speculated to be an antidote for opium.

Other Uses

- Bait: John C Duval reported in 1936 that the odor of asafoetida is attractive to the wolf, a matter of common knowledge, he says, along the Texas/Mexico border. It is also used as one of several possible scent baits, most notably for catfish and pike.
- May also be used as a moth (Lepidoptera) light trap attractant by collectors—when mixed by approximately 1 part to 3 parts with a sweet, fruit jelly.
- Repelling spirits: In Jamaica, asafoetida is traditionally applied to a baby's anterior fontanel (Jamaican patois *mole*) to prevent spirits (Jamaican patois *duppies*) from entering the baby through the fontanel. In the African-American Hoodoo tradition, asafoetida is used in magic spells, as it is believed to have the power both to protect and to curse.
- In ceremonial magick, especially from *The Key of Solomon the King*, it is used to protect the magus from daemonic forces and to evoke the same and bind them.

History in the West

It was familiar in the early Mediterranean, having come by land across Iran. Though it is generally forgotten now in Europe, it is still widely used in India. It emerged into Europe from a conquering expedition of Alexander the Great, who, after returning from a trip to northeastern Persia, thought they had found a plant almost identical to the famed silphium of Cyrene in North Africa—though less tasty.

Dioscorides, in the first century, wrote, "the Cyrenaic kind, even if one just tastes it, at once arouses a humour throughout the body and has a very healthy aroma, so that it is not noticed on the breath, or only a little; but the Median [Iranian] is weaker in power and has a nastier smell." Nevertheless, it could be substituted for silphium in cooking, which was fortunate, because a few decades after Dioscorides's time, the true silphium of Cyrene became extinct, and asafoetida became more popular amongst physicians, as well as cooks.

Asafoetida is also mentioned multiple times in Jewish literature, such as the Mishnah. Maimonides also writes in the Mishneh Torah "In the rainy season, one should eat warm food with much spice, but a limited amount of mustard and asafoetida." Asafoetida was described by a number of Arab and Islamic scientists and pharmacists. Avicenna discussed the effects of asafoetida on digestion. Ibn al-Baitar and Fakhr al-Din al-Razi described some positive medicinal affects of it on the respiratory system.

After the Roman Empire fell, until the 16th century, asafoetida was rare in Europe, and if ever encountered, it was viewed as a medicine. "If used in cookery, it would ruin every dish because of its dreadful smell," asserted García de Orta's European guest. Nonsense, García replied, "nothing is more widely used in every part of India, both in medicine and in cookery. All the Hindus who can afford it buy it to add to their food."

Cultivation and Manufacture

The resin-like gum comes from the dried sap extracted from the stem and roots and is used as a spice. The resin is greyish-white when fresh but dries to a dark amber colour. The asafoetida resin is difficult to grate and is traditionally crushed between stones or with a hammer. Today, the most commonly available form is compounded asafoetida, a fine powder containing 30% asafoetida resin, along with rice flour and gum arabic.

Ferula assafoetida is a monoecious, herbaceous, perennial plant of the family Umbelliferae, also called Apiaceae. It grows to 2 metres high, with a circular mass of 30–40 cm leaves. Stem leaves have wide sheathing petioles. Flowering stems are 2.5–3 metres high and 10 cm thick and hollow, with a number of schizogenous ducts in the cortex containing the resinous gum. Flowers are pale greenish yellow produced in large compound umbels. Fruits are oval, flat, thin, reddish brown and have a milky juice. Roots are thick, massive, and pulpy. They yield a resin similar to that of the stems. All parts of the plant have the distinctive fetid smell.

Composition

Typical asafoetida contains about 40–64% resin, 25% endogeneous gum, 10–17% volatile oil, and 1.5–10% ash. The resin portion is known to contain asaresinotannols 'A' and 'B', ferulic acid, umbelliferone and four unidentified compounds.

Bibliography

Arguello D. S.: *Production, Post-harvest Management and Exportation of Tropical Fruits in Costa Rica,* Wageningen, Technical Centre for Agricultural and Rural Cooperation, 1992.

Arora, Chandralekha : *Advances in Horticulture Science Research,* Oxford Book Company, Delhi, 2009.

Barbeau G.: *Tropical Fruits in Nicaragua,* Managua, Nicaragua Ministerio de Desarrollo Agropecuario, Agraria, 1990.

Bhattacharjee, Supriya Kumar : *Amenity Horticulture, Biotechnology and Post-Harvest Technology,* Pointer, Delhi, 2006.

Chadha K. L. and Pareek O. P.: *Advances in Horticulture: Fruit Crops,* New Delhi, Malhotra Publishing House, 1993.

Crucefix D.: *Avocado Variety Selection for Export Development,* Roseau, CARDI, 1996.

Currah L. and Proctor F. J.: *Onions in Tropical Regions,* Kent, Natural Resources Institute, 1990.

Dhatt A. S. and Singh Z.: *Propagation and Rootstocks of Citrus,* New Delhi, Malhotra Publishers, 1993.

Doijode S. D.: *Seed Germination in Fruits,* New Delhi, Malhotra Publishers, 1993.

Featherly H. I.: *Taxonomic Terminology of the Higher Plants,* USA, Iowa State College Press, 1954.

Gibbs, D.R. : *Chemotaxonomy of Flowering Plants,* McGill-Queen's University Press, Montreal, 1974.

Glowinski Louis: *The Complete Book of Fruit Growing in Australia,* Australia, Lothian Publishing Company Pty. Ltd., 1991.

Gunjate R. T. and Tawde A. B.: *Propagation and Rootstocks of Mango,* New Delhi, Malhotra, 1993.

Harlan, Jack R.: *The Living Fields: Our Agricultural Heritage*, Cambridge: Cambridge University Press, 1998.

Herklots G. A. C.: *Vegetables in South East Asia*, London, George Allen & Unwin Ltd., 1972.

Jeffers P.: *Evaluation of Four Onion Varieties in Montserrat*, Plymouth, CARDI, 1992.

Jerram, M.R.K. : *A Text-Book on Forest Management : In Agriculture, Horticulture and Forestry*, Asiatic Pub, Delhi, 2006.

Jugale, V.B. : *Horticulture Economy of Maharashtra*, Shruti, Delhi, 2004.

Malins A.: *Postharvest Handling of Pineapple and Mango*, Port of Spain, Trinidad and Tobago, 1992.

Matsuoka H.: *Cultivation of Panicum Genetic Resources for Evaluation of Characteristics*, Tokyo, JICA, 1997.

Mitra S.: *Postharvest Physiology and Storage of Tropical and Subtropical Fruits*, Oxon, CABI, 1997.

Morton Julia F.: *Fruits of Warm Climates*, Miami, Julia F. Morton Publisher, 1987.

Oldham P.: *Cost of Production of Major Tree Crops in Dominica*, Roseau, Ministry of Agriculture, 1991.

Peter, K.V. : *Basics of Horticulture*, New India, Delhi, 2009.

Pilgrim R.: *Post Harvest Handling of Minor Exotics*, St. George's, Grenada, 1996.

Prasad, Surendra : *Agro's Dictionary of Horticulture*, Agrobios, Delhi, 2003.

Sharma, V.K. : *Advances in Horticulture : Strategies, Production, Plant Protection and Value Addition*, Deep and Deep, Delhi, 2004.

Singh H. P. and Chadha K. L.: *Genetic Resources of Citrus*, New Delhi, Malhotra Publishing House, 1993.

Singh, D.K. : *Elements of Horticulture*, Agrotech, Delhi, 2005.

Singh, H.P. : *Challenges of Climate Change - Indian Horticulture*, Westville Publishing House, Delhi, 2010.

Singh, S.P. : *Advances in Horticulture and Forestry*, Scientific, Delhi, 2001.

Thavarasook C.: *Proceedings of the Mungbean Meeting 90, Chiang Mai, Thailand,* Bangkok, Tropical Agricultural Research Centre, 1991.

Weiss, R.F. : *Herbal Medicine*, Beaconsfield Arcanum, Beaconsfield, 1991 .

Whitwell A.: *Dominica Orchard Crop Management and Research Project: Pest and Disease Management,* United Kingdom, Natural Resources Institute, 1991.

Index

❑❑❑